Climate-Smart Food

Dave Reay

Climate-Smart Food

palgrave
macmillan

Dave Reay
School of Geosciences
University of Edinburgh
Edinburgh, UK

ISBN 978-3-030-18205-2   ISBN 978-3-030-18206-9   (eBook)
https://doi.org/10.1007/978-3-030-18206-9

This Palgrave Pivot imprint is published by the registered company Springer Nature
Switzerland AG
The registered company address is: Gewerbestrasse 11, 6330 Cham, Switzerland

To Sarah, you're amazing

PREFACE

Food is a wonderful thing. At times, a simple fuel, at others, the key binding agent for families, communities, even whole cultures. It anchors our memories too. My youngest daughter, Molly, can unerringly remember people and places that the rest of us have long forgotten, each recollection planted in her mind by whatever we were eating at the time.

My own childhood was similarly food obsessed. Money was tight, and with three hungry siblings, the competition at the dinner table was fierce. Every mouthful was watched, any leftovers devoured. Some meals had strictly adhered-to ceremonies and rules. If there was pudding (the sight of spoons on the table always made us happy), then was it custard? If so, whose turn was it to have the skin? If we had a Sunday roast, then the ante would be upped even further as we vied to be the one chosen for 'the bone'.

My little sister Elisabeth always seemed to get the nod for these, and I distinctly remember dinner times with her sitting across the table grappling a giant pork thigh bone in her tiny grease-covered hands and systematically gnawing it bare. Fairness was paramount and outrage was the response if anyone dared to cheat. At primary school, I had the misfortune of my big brother, Mike, being table monitor for lunches. This lofty appointment included control of the small squares of chocolate to go with our semolina. Little chocolate ever came my way.

By adolescence, I was a frequent and highly experimental cook. Returning home ravenous each day, cupboards would be scoured for ingredients and the results were unpredictable—even my bottomless 15-year-old self could not stomach pork-lard doughnuts. As growing kids, we may have been hungry a lot of the time, but our food delights were many. A friend's dad worked at a cake factory, and at most school lunch-times, we could sneak into the workers' cloakroom and feast on piles of still-warm offcuts. At home, any visit by my uncle Bern was always met with whoops of delight as he produced box upon box of Kipling cakes for tea, then cooked up breakfast bacon sandwiches so mouth-wateringly gorgeous that they could even entice teenage boys out of bed on a Sunday morning.

Food has always been at the centre of my professional life too, the focus being on its interactions with climate change. From measuring nitrous oxide in farm drainage ditches, through inventing methane sponges for cattle sheds, to investigating the climate promise and pitfalls of food waste and dietary change, it has always intrigued me.

Ultimately, food is personal, and so this book is too. It has opened my eyes to the fragility of the banana in my lunch box, of the daily miracle that is coffee, and the harsh realities of mass-produced chicken. Most shocking were the climate challenges already being faced by smallholder farmers around the world and the huge amounts of the precious food they produce that we then waste.

In trying to understand the climate past, present and future of a single day's food, I have drawn from many hundreds of scientific studies, government reports and media articles. Thank you to this multitude of authors, and to Google, for helping me to find your work (how else could I have discovered gems like potato 'shatter risk'?). Two truly outstanding resources that I have relied on throughout have been CarbonBrief.org and OurWorldinData.org—if you want well-researched and brilliantly accessible information on climate change, food and a host of other fundamental facets of civilisation, then these are a must. Likewise, the excellent 'Food & Climate Research Network' (FCRN.org.uk) has been invaluable in finding well-buried data and information on emissions from food.

Big thanks to the University of Edinburgh, and my own School of Geosciences, for supporting me in writing this book, and especially for allowing it to be open access worldwide. Thanks also to the starry array of friends who I am so blessed to know, including everyone in Threemiletown (chuck another veggie sausage on the BBQ for me Mahmoud!), Mel, Ceri and Sandy McEwan, Erika Warnatzsch, Sam Metaxas, Stephen Porter, Andi Moring, Pete Higgins, Meredith Corey, all the 'Carbon Masters' and so many more, at Edinburgh University and beyond. Thanks too to folk like David Attenborough, Keith Smith, Pete Smith, Katharine Hayhoe,

Michael Mann and Greta Thunberg, who continue to inspire me on all things environment and climate.

Thanks to Martial Bernoux and Aziz Elbehri at the United Nations Food and Agriculture Organization for planting the seed for the idea of this book, and for my editor at Palgrave, Rachael Ballard, for believing in it. Special thanks to my friend and utterly brilliant colleague Hannah Ritchie, who provided the *Our World in Data* global food maps that appear throughout this book. Thanks to Mike, Paul and Elisabeth for so many warm (and some incendiary) dinnertime memories, and to Mum and Dad for putting food on the table whatever the odds. Thanks to my late uncle Bern too, for wonderful food memories too numerous to count. Finally, thank you to Sarah, Maddy, Molly and Ginny. Writing this book would have been impossible without your unfailing support and love. You mean the world to me.

Edinburgh, UK   Dave Reay

CONTENTS

LIST OF FIGURES

CHAPTER 1

Introduction

Abstract Climate change poses a severe and growing threat to food security around the world. Our food is also a major driver of climate change. Here we provide an overview of these intertwined global challenges and the current state of progress (or lack thereof) in addressing them. We introduce the concept of climate-smart food, whereby climate resilience and productivity are increased while greenhouse gas emissions are simultaneously reduced. Finally, we map out the specific foods to be explored in-depth, from farm, vineyard or ocean to Scottish dinner table.

Keywords Paris Agreement • 1.5 degrees • Food security • Food waste • Malnutrition • Carbon footprint • Food miles

Time is against us. The world has already warmed by an average of 1 degree Celsius, as decades of rising greenhouse gas emissions have accumulated in our atmosphere. Devastating impacts are predicted if we fail to hold average warming well below 2 degrees Celsius this century. Any delay in tackling climate change, even one that allows the seemingly minor upward creep in the mercury from 1.5 to 2 degrees Celsius, will intensify droughts and floods, expose hundreds of millions more people to heat waves and risk complete destruction of the world's tropical coral reefs [1]. Sleepwalk into the steeper twenty-first-century warming pathways of 3, 4

© The Author(s) 2019
D. Reay, *Climate-Smart Food*,
https://doi.org/10.1007/978-3-030-18206-9_1

or even 5 degrees Celsius and the climate change threats becomes existential to civilisation itself [2].

The Paris Climate Agreement—a framework in which all nations can commit to and then implement climate action [3]—is the best game in town for steering us away from such Hollywood-fodder futures. So far though, the political rhetoric on urgent action does not match reality.

Current Paris commitments would still mean warming of about 3 degrees Celsius, and we are fast eating through our remaining 'safe' emissions budget [4]; metaphorically, in continued fossil fuel burning, and quite literally, through our food. Each carrot and tomato, burger and chicken drumstick, every food has a carbon footprint and feeding us all requires an awful lot of it. Even if (and it's still a big if) we manage to radically cut global fossil fuel use in the next decade, rising emissions from agriculture could slam the door shut on our chances for a safer climate future.

The world's food system is now responsible for over a quarter of greenhouse gas emissions [5]. Population is set to rise to around ten billion by the middle of the century at the same time as droughts, floods, heat waves and disease increasingly threaten food security. Feeding everyone well without blowing the climate budget represents one of the biggest challenges our society has ever faced. Signs are we're not match fit.

One in nine people alive today don't have enough to eat, while two billion of us consume too much [6]. Western diets have become much more calorie and meat intensive [7], ramping up emissions and damaging the health of both humans and the planetary systems we all depend on [6]. At the same time a billion people are lacking enough protein, one-third of children under five are stunted and some two billion people suffer from micronutrient deficiencies [8]. Tragically, around a quarter of all the food produced for human consumption doesn't even get eaten [9]. At an annual cost of nearly $1 trillion, global food loss and waste accounts for an estimated 8 per cent of total greenhouse gas emissions—if wasted food were a country it would come behind only China and the US in the list of biggest emitters on the planet [10].

Over the last decade the idea of Climate-Smart Agriculture—where these two Hitchcockian birds of food insecurity and climate change are hit with one stone—has grown apace. Led by the United Nations' Food & Agriculture Organisation it has developed from a few small-scale pilot farms into a global powerhouse of research, capacity-building and sharing of good practice on how food systems can become more productive, more resilient to climate change *and* lower carbon all at the same time [11].

Sitting hungrily at the receiving end of those food systems are the consumers: us. We are each connected to hundreds, maybe thousands, of other people and places through what we eat every day. We are connected to their soil, water and climate too. This book traces just a few threads in the tangled global web that is food and climate change—those of one day's food and drink for my family. Today's menu, from breakfast through to dinner, is a special one:

Climate-Smart Food

Breakfast
Orange Juice
Toast
Tea & Milk

Break Time
Chocolate Bar
Banana
Coffee

Lunch
Chicken Curry and Rice
Bag of Nachos

Dinner
Fish and Chips
Champagne

Some things on the menu—like tea and coffee, bread and chocolate—are regulars for most of us. Some, especially the expensive champagne with

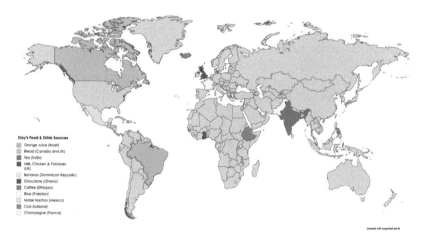

Fig. 1.1   Source countries for our day's food (created using mapchart.net)

dinner, are a very rare treat. Between them their stories span five continents and many nations. They are testament to the global nature of what we consume and its global exposure to the impacts of climate change. Unless your name is Lily Bollinger, some of your own daily food and drink will undoubtedly be different. Your tea might come from China instead of India, and your orange juice may be American not Brazilian. For milk or chicken, you may well have already made the switch to plant- and fungi-based substitutes. As such, the precise carbon footprints[1] will vary and the best climate-smart responses change (as we will see, local context is king) (Fig. 1.1).

The 13 foods and drinks explored here are inevitably a tiny Scotland-centred snap shot of the behemoth that is our global food system and its tumultuous relationship with climate. Each is followed back from our scuffed West Lothian dinner table to the field, barn or ocean waters that it

---

[1] Here, and throughout the book, the 'carbon footprint' is the amount of greenhouse gas emitted per unit of a particular food—the emissions per tonne of wheat, for example, or per glass of orange juice. Unless otherwise stated, all figures stated represent emissions of 'carbon dioxide equivalents' (or CO_2e). The CO_2e metric includes and standardises emissions of non-carbon dioxide greenhouse gases, like methane and nitrous oxide. It accounts for their differing lifetimes in the atmosphere and different 'radiative forcing' (warming) properties so, for example, 1 tonne of nitrous oxide emissions has about 300 times the warming effect of 1 tonne of carbon dioxide, and so would appear as 300 tonnes of CO_2e.

originally came from. We'll explore their carbon footprints, the extreme weather events they have endured and the climate threats they face in the coming decades. Crucially, we'll look at how climate-smart solutions could alter this future: whether such magic win-win-wins for food security, livelihoods and climate change even exist and, if so, whether we as consumers can help deliver them. Let's see.

REFERENCES

1. Carbonbrief.org. *In-Depth Q&A: The IPCC's Special Report on Climate Change at 1.5C.* https://www.carbonbrief.org/in-depth-qa-ipccs-special-report-on-climate-change-at-one-point-five-c (2018).
2. IPCC. *Climate Change 2013: The Physical Science Basis. Contribution of Working Group I to the Fifth Assessment Report of the Intergovernmental Panel on Climate Change* (Cambridge University Press, Cambridge, UK and New York, NY, 2014).
3. Carbonbrief.org. *The Paris Agreement on Climate Change.* https://www.carbonbrief.org/interactive-the-paris-agreement-on-climate-change (2015).
4. Carbonbrief.org. *Analysis: Why the IPCC 1.5C Report Expanded the Carbon Budget.* https://www.carbonbrief.org/analysis-why-the-ipcc-1-5c-report-expanded-the-carbon-budget (2018).
5. Carbonbrief.org. *Failure to Tackle Food Demand Could Make 1.5C Limit Unachievable.* https://www.carbonbrief.org/guest-post-failure-to-tackle-food-demand-could-make-1-point-5-c-limit-unachievable (2016).
6. Willett, W. *et al.* Food in the Anthropocene: The EAT—Lancet Commission on healthy diets from sustainable food systems. *Lancet* **393**, 447–49P (2019).
7. Kearney, J. Food consumption trends and drivers. *Philos. Trans. R. Soc., B: Biol. Sci.* **365**, 2793–2807 (2010).
8. Ritchie, H., Reay, D. S. & Higgins, P. Beyond calories: A holistic assessment of the global food system. *Front. Sustain. Food Syst.* **2**, 57 (2018).
9. Silva, J. G. d. *Food Losses and Waste: A Challenge to Sustainable Development.* http://www.fao.org/save-food/news-and-multimedia/news/news-details/en/c/429182/ (2016).
10. Hanson, C. & Mitchell, P. The business case for reducing food loss and waste. *Champions* **12**, 7–8 (2017).
11. Rosenstock, T. S., Nowak, A. & Girvetz, E. *The Climate-Smart Agriculture Papers Investigating the Business of a Productive, Resilient and Low Emission Future* (Springer, 2019).

PART I

Breakfast

CHAPTER 2

Climate-Smart Orange Juice

Abstract Our daily glass of orange juice has travelled a long way to the breakfast table. Travel is part of its carbon footprint, but growing the oranges in the first place dominates emissions. Each glass has a total footprint of around 200 grams. In the UK we waste approximately 50,000 tonnes of orange juice each year—reducing household waste and improving the efficiency of water and fertiliser use on farms stand out as ways to cut the carbon footprint of orange juice. Growers in Brazil and the US are battling citrus greening disease and those in Florida have been devastated by frost damage in the past. In the future, climate change is set to bring greater pest and disease risks alongside drought and heat-stress issues. Strategies such as irrigation, soil moisture management and biological pest control all emerge as potentially powerful climate-smart solutions, but for some orange farmers abandoning orange growing altogether may be the only long-term answer.

Keywords Oranges • Citrus • Brazil • US • São Paulo • Drought • Citrus greening • Irrigation • Carbon footprint • Resilience • Amazon

Scotland is famous for many things, including its food. Some fruits grow superbly well here, with our own small plot of raspberry canes groaning under the weight of berries each year. The chill summer rain sweeping past the window is definitely not citrus-growing weather though. Orange trees

© The Author(s) 2019
D. Reay, *Climate-Smart Food*,
https://doi.org/10.1007/978-3-030-18206-9_2

need plenty of sunshine and warmth—temperatures below 7 degrees Celsius will often kill them. As such, the breakfast orange juice enjoyed by millions may have travelled a very long way.

Brazil [1] is now the world's biggest producer of orange juice, with the US the only other big league player. The near-perfect growing conditions found in São Paulo and Florida mean that just these two regions are together responsible for around 80 per cent of all production.

Since the 1960s the rise of Brazil as a global orange-growing super-power has been unstoppable. Just as global demand was growing fast, production in the usually sun-drenched groves of Florida began to falter. Extreme weather was the culprit. The odd frost, even in Florida, is not that unusual, but in the 1960s, 1970s and 1980s, the sunshine state was hit with a series of record-breaking cold snaps that left acres of blackened mush-filled oranges hanging in its wake. The 1980s' frosts were cata-strophic. An initial wave of hard frosts in 1981 had already caused a lot of damage in northern Florida, with costs at the time estimated at over $1 billion. Then, in 1983, came the so-called Freeze of the Century. Two days of lethal cold devastated orange production right across the state, with up to 90 per cent of trees and fruit damaged [2]. Further damaging frost events in 1985 and 1989 ensured the Florida orange industry could never really recover the ground lost to its frost-free Brazilian competitors. Today Brazil produces more than double (16 million tonnes a year) the amount of oranges grown in the US (Fig. 2.1).

Like most types of food, the carbon footprint of orange juice is domi-nated by on-farm, or in this case on-grove, emissions. Most are a result of the nitrogen fertilisers that are applied to improve tree growth (and the nitrous oxide—a powerful greenhouse gas—that is then emitted from the soils). Others arise from the energy required for fertiliser, herbicide and pesticide production, and from the fuel used to power harvesting machin-ery [4]. By the time the oranges trundle through the farm gate, up to 60 per cent of their life-cycle footprint is already invisibly embedded in their juicy flesh [5].

What happens next can still have important implications for the lifetime carbon footprint of the juice we drink. First, the truckloads of harvested oranges are driven to nearby processing centres, adding a small extra slice of emissions on the way. Next comes quality checking and sorting, before the oranges are washed and squeezed to produce super-fresh raw juice [6]. If destined for foreign shores it's either sent direct to port in refrigerated tankers or is first concentrated—concentrating the juice in its home

Orange production, 2014
Annual agricultural production of oranges, measured in tonnes per year.

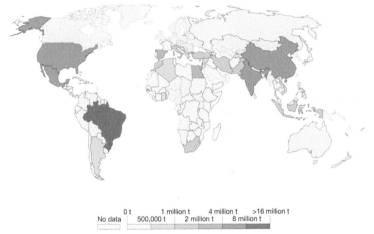

| No data | 0 t | 1 million t | 4 million t | >16 million t |
| | 500,000 t | 2 million t | 8 million t | |

Source: UN Food and Agriculture Organization (FAO)   CC BY

Fig. 2.1 Global orange production in 2014 by country of origin (Source: Our World in Data) [3]. Available at: https://ourworldindata.org/grapher/orange-production

country radically reduces its shipping weight (and so transport emissions), with the extracted water then being replaced once it reaches its destination. Finally, in its country of consumption, the juice is given a big top-up of water if required, packaged-up and sent to our stores.

For major orange juice importers like the UK, the overall carbon footprint ranges from the equivalent of 100 grams of carbon dioxide per serving for 'from concentrate' juice up to 400 grams for the more expensive types of pure juice [7]. Our own cheap and cheerful economy carton [8] is therefore down at the lower end of this range. As the bulk of emissions from orange juice arise before we even get it home, the potential for consumers to make it more climate-smart would seem limited. There is, however, one crucial way in which every juice-drinking household can send a low carbon ripple right back through the global supply chain: avoiding waste.

Robust data on just how much orange juice is wasted by households is scarce. One of the best available sources—and one that I'll use frequently in the coming chapters—is the UK's Waste & Resources Action Programme or WRAP [9]. Their reports are based on household surveys and cover many different foods and drinks. Information on orange juice is lumped together with other juices in the Fruit Juice category, so includes apple, pineapple and others. As over half of the fruit juice drunk in the UK is orange [10] we can assume that about half of the overall juice wastage is also orange. The amounts this implies are startling. Of the 1.1 million tonnes of all types of juice purchased by UK households in 2012, more than one-tenth was wasted. All was deemed avoidable [9].

For the UK alone this means over 50,000 tonnes of orange juice, produced on farms around the world and shipped thousands of miles, ends up down the drain. Even at the low-end carbon footprint of 100 grams per glass the avoidable emissions from this waste total over 25,000 tonnes. The main reasons cited for it not getting drunk are either that the juice isn't used in time or that too much is served. Through waste-aware shopping and refrigerator habits, and fewer vat-sized servings, we can therefore slash the overall carbon footprint of our daily juice. To find out if this could be cut even further, and just how vulnerable global orange production is to climate change, means going right back to the starting point of our juice's long journey by land and sea: to sunny São Paulo in Brazil.

* * *

In terms of an accidental climate-smart response, one could argue that 'the market' has delivered on orange juice. Frost-prone orange production in Florida has shifted to the outwardly more reliable climate envelope of São Paulo, so giving greater resilience to extreme weather events. Frost, however, is not the only severe weather risk stalking orange groves.

Growing oranges requires a plentiful water supply—the average orange tree will drink its way through around 30 gallons a day during summer [11]. In Brazil's orange-growing powerhouse state of São Paulo heavy rains usually sweep in between October and March each year. This reliable wet season has meant few farmers use irrigation, instead they rely on the summer rains steadily drenching their thirsty soils. Sometimes, though, the rains fail. In the Brazilian summer of 2013–14 the moist air that would normally deliver massive volumes of rain to São Paulo state was blocked by a stubborn high pressure system [12]. Soils began to dry and lake levels to drop. By early September 19 cities were in the grip of water rationing, hydroelectric plants were struggling to generate enough energy and states began arguing over access to the remaining water supplies. For farmers, including São Paulo's 10,000-strong army of orange growers, the drought impacts were already obvious, yet hopes were high that the following summer rains would relieve any water supply issues. Then those rains failed too [13].

As the 2015 summer progressed, temperatures soared and the worst drought on record unfolded. By March some desperate residents in São Paulo city had started drilling through basement floors to try and find groundwater [14]. Hoarding of water in buckets and cans provided the perfect breeding ground for mosquitoes, with the incidence of dengue fever soaring [15]. Out in the parched fields of São Paulo state, the authorities locked up taps normally used to pump water to farms. Many farmers could do little more than watch their crops wither before their eyes. Around one-fifth of the entire state's citrus crop was destroyed [16].

The rains did return, eventually. A powerful El Niño (a reversal of wind and ocean currents in the Pacific) ensured that by late 2015 the streams and rivers of São Paulo state could flow again. As reservoirs refilled and water rationing was lifted, life returned to normal. Yet concerns remained that the impacts of future droughts could be even worse and that opportunities to increase resilience—such as improved water storage and reduced leakage—might be missed [15].

Rapid population growth and soaring water demand were deemed to be the primary reasons for the drought being so very severe. São Paulo city itself is now home to over 20 million people, and water demand in the city is estimated to have risen fivefold since the 1960s [17]. Whether climate change will exacerbate water security problems even further remains unclear. An increase in severe weather events, including droughts, is expected for Brazil as whole [1] and it is feared that further deforestation in the Amazon will rob southern states of the moisture-rich air masses that the tropical forests generate [18]. In São Paulo state specifically, a trend of more intense rainfall events alongside more dry days is expected [19], with higher temperatures meaning any rain that does fall evaporates more quickly. For orange farmers, in particular, this testing climate future may have an extra sting in the tail.

Direct impacts of severe weather on farms—drought-shrivelled crops, for example—can be all too obvious. Yet our changing climate may also affect the myriad pests and diseases that plague food production. The number one disease threat to orange trees around the world goes by the tongue-tangling name of Huánglóngbìng (try saying that after a few vodka and oranges). More commonly known as 'citrus greening' or simply HB, it is caused by a bacterium passed from tree to tree by tiny leafhopper insects. Early signs of infection are poor tree growth, blotchy foliage and misshapen green fruit that taste bitter and drop early [20].

As the bacteria spreads further, it starves the whole tree of nutrients and eventually kills it. The disease is now widespread across Asia, Africa and the Americas. For the orange-growing powerhouses of São Paulo and Florida it is a disaster. Between 2006 and 2011, citrus greening in Florida alone is estimated to have cost over $4 billion and put 8,000 people out of work [21]. Inevitably, the prices of our breakfast orange juice have also surged as supplies from Brazil and Florida have faltered [22].

There is still no effective treatment for citrus greening, with control efforts instead focused on killing the leafhopper insects that transmit it. Climate change could make control of such fruit tree diseases even harder, as conditions become more favourable for the mites and insects that carry them. In São Paulo, many farmers are already dealing with swathes of wilting orange trees suffering from variegated chlorosis, another disease carried by leafhoppers [23]. Warmer growing seasons in the future may increase the production of young tree shoots and so boost the populations of leafhopper insects that live on them. Were such disease outbreaks to hit orange groves already weakened by droughts and heat waves the impacts would be devastating [24].

With so many challenges the future of our budget-store orange juice appears rather bleak. Certainly, prices will rise further if global supplies are increasingly squeezed by severe weather impacts and disease spread. Following a run of poor harvests and uncertain profits over the last decade, some farmers have grubbed up their orange groves entirely [22, 25].

Most are not giving up yet though. Instead they are altering growing strategies, using new technologies and data, and drawing more and more on expert advice and support. Across the industry, the global nature of the orange juice supply chain is helping to highlight vulnerabilities and strengthen responses. Slowly but surely, orange juice is getting climate-smart.

* * *

Making orange production more resilient in a changing climate inevitably means addressing drought risks. As rainfall becomes more unreliable, so irrigation becomes a mainstay of adaptation. For citrus growers in arid areas irrigation is already common practice, but even in wetter areas like Florida the big water demands of orange trees during summer mean most farmers can't rely on rainfall alone [26]. Exactly how and when Florida's farmers use irrigation has had to evolve over time as concerns over wider water scarcity in the state have grown. Instead of large overhead sprinklers—which often wasted water and increased the risk of frost damage to fruit—drip and micro-sprinklers are now employed [27]. These networks of narrow tubes usually run over the surface of the soil and deliver water through an array of small holes or spray heads.

Traditionally, the frequency of irrigation would be based around standard growing calendars or as a response to symptoms of water stress. But by the time signs of drought are evident in the trees much of the damage may already have been done [28]. The advent of computer-controlled automation and smart soil moisture sensors means precision irrigation is now possible. By constantly monitoring water availability and use these systems can automatically increase or decrease flows. Some are even linked up to weather stations so they can make real-time adjustments for extra water losses on hot or windy days [29].

In this way, overall water use can be radically reduced, and loss of fertilisers and pesticides from over-watered soils minimised too—leaching and run-off from farmland is a big issue for water quality in many areas. Importantly, use of such smart micro-irrigation can boost orange production even as rainfall becomes more unreliable. It allows farmers to deliver water just where and when it is needed, and to promote deep root growth in the trees (deeper roots then meaning the trees are less prone to drought stress in the future).

For orange juice's carbon footprint, the benefits of smart irrigation can also be big. Following recent droughts, growers in Brazil have been copying their US counterparts and making increasing use of irrigation to boost yields [30]. Given that the 2015 drought destroyed around one-fifth of the orange crop in São Paulo [31], effective irrigation during such droughts in the future could save the equivalent of up to a quarter of a billion litres of juice, and so an impressive twenty-thousand tonnes or so of carbon dioxide emissions along with it.

Irrigation may be a core part of climate-smart orange juice then, delivering higher yields and increased drought resilience and avoiding the greenhouse gas emissions associated with lost fruit and trees. It does of course rely on their being enough water available. If supplies are cut off—as happened in the big São Paulo drought—then no amount of fancy micro-irrigation kits will keep the farms from harm. Instead, integrating the water needs of farms with state-wide drought plans can highlight supply risks. Many orange farms in Florida already make use of reclaimed water (water derived from waste treatment [32]) to buffer them against droughts and water supply restrictions. Others have made more use of on-site rainwater harvesting and mulch their soils to cut water losses from evaporation too.

Keeping diseases like citrus greening at bay is proving more difficult. With no effective cure, many farmers resort to burning infected trees in the hope of stopping its spread. Globally an estimated 100 million trees have now been destroyed by the disease [33]. As the average orange tree locks up 100 kilograms of carbon in its stem and branches [34] these losses may have big implications for the atmosphere and our climate, as well as for the world's orange farmers. Intensive pesticide use can limit the disease-carrying leafhoppers, but with it come risks to water quality, biodiversity and human health [35]. Truly climate-smart orange production therefore means protecting groves from further destruction in a way that also avoids increased pollution of air, soil and water. In the Caribbean some orange farmers are doing exactly that.

Two national programmes, in Jamaica and Belize, scoped out the options for tackling citrus greening that would work best for their own farmers. They first set up area-wide management programmes that allowed farmers to share best practices, to get expert training and to ensure control efforts were coordinated. The systems for testing for infection were also bolstered. Biological pest control—using natural leafhopper predators and diseases—was then used instead of insecticides. New plant nurseries were set up that could provide disease-free orange saplings and existing trees were made more resilient by improving their nutrition and controlling grove weeds. Within just two years the farmers were reporting higher yields and better quality oranges [33].

Disease, drought and see-sawing prices will still force many growers to diversify what they grow or opt out of oranges altogether. Yet, the early signs of success from the Caribbean programmes show just how effective coordinated action that directly involves farmers can be. With ever-evolving technologies for water management, improving disease control, and a real push for local training and support, the orange growers of São Paulo (and our breakfast orange juice) might just be OK.

REFERENCES

1. Timperley, J. The Carbon Brief Profile: Brazil. *Carbonbrief.org.* https://www.carbonbrief.org/the-carbon-brief-profile-brazil (2018).
2. Nordheimer, J. 'Freeze of the century' damages 90% of the citrus crop in Florida. *New York Times.* https://www.nytimes.com/1985/01/23/us/freeze-of-the-century-damages-90-of-the-citrus-crop-in-florida.html (1985).
3. Ritchie, H. Global orange production, 2014. *OurWorldinData.org.* https://ourworldindata.org/grapher/orange-production (2018).
4. Mordini, M. *et al. Carbon & Water Footprint of Orange and Strawberries.* Federal Department of Economic Affairs, Zurich, Switzerland. http://citeseerx.ist.psu.edu/viewdoc/download?doi=10.1.1.690.8098&rep=rep1&type=pdf (2009)
5. Spreen, T., Dwivedi, P. & Goodrich-Schneider, R. Estimating the carbon footprint of Florida orange juice. *Proc. Food Syst. Dyn.* 95–101 (2010).
6. Plummer, R. Brazil's orange juice. *BBC Online.* http://news.bbc.co.uk/1/shared/spl/hi/picture_gallery/06/business_brazil0s_orange_juice/html/1.stm (2009).
7. Tesco. Product Carbon Footprint summary. https://www.tescoplc.com/assets/files/cms/Tesco_Product_Carbon_Footprints_Summary(1).pdf (2012).
8. LoveFood.com. *Basic Britain: Orange Juice.* https://www.lovefood.com/news/58618/basic-britain-orange-juice (2012).
9. WRAP. Household food and drink waste in the United Kingdom 2012. *Waste and Resource Action Programme.* http://www.wrap.org.uk/sites/files/wrap/hhfdw-2012-main.pdf.pdf (2013).
10. Association, B. S. D. The UK Soft Drinks Annual Report 2015. http://www.britishsoftdrinks.com/write/mediauploads/publications/bsda_annual_report_2015.pdf (2015).
11. Wright, G. C. *Irrigating Citrus Trees.* University of Arizona Extension. https://extension.arizona.edu/sites/extension.arizona.edu/files/pubs/az1151.pdf (2000).
12. Nobre, C. A. & Marengo, J. A. Water crises and megacities in Brazil: Meteorological context of the São Paulo drought of 2014–2015. *Global Water Forum.* http://www.globalwaterforum.org/2016/10/17/water-crises-and-megacities-in-brazil-meteorological-context-of-the-sao-paulo-drought-of-2014-2015/ (2016).
13. Watts, J. Brazil drought crisis leads to rationing and tensions. *The Guardian.* https://www.theguardian.com/weather/2014/sep/05/brazil-drought-crisis-rationing (2014).

14. Carvalho, L. Megacity drought: Sao Paulo withers after dry 'wet season'. *The Conversation.* https://theconversation.com/megacity-drought-sao-paulo-withers-after-dry-wet-season-42799 (2015).
15. Stauffer, C. Drought ends in Brazil's Sao Paulo but future still uncertain. *Reuters.* https://uk.reuters.com/article/us-brazil-water/drought-ends-in-brazils-sao-paulo-but-future-still-uncertain-idUKKCN0VR1YJ (2016).
16. Brito, R. P. d., Miguel, P. L. d. S. & Pereira, S. C. *Academy of Management Proceedings,* 15676 (Academy of Management, Briarcliff Manor, NY 10510).
17. McSweeney, R. Climate change 'not a major influence' on Brazil drought, study says. *Carbonbrief.org.* https://www.carbonbrief.org/climate-change-not-a-major-influence-on-brazil-drought-study-says (2015).
18. Watts, J. The Amazon effect: How deforestation is starving São Paulo of water. *The Guardian.* https://www.theguardian.com/cities/2017/nov/28/sao-paulo-water-amazon-deforestation (2017).
19. Cavalcanti, I. F. *et al.* Projections of precipitation changes in two vulnerable regions of São Paulo state, Brazil. *Am. J. Clim. Change* **6,** 268 (2017).
20. Moffis, B. L., Burrow, J. D., Dewdney, M. M. & Rogers, M. E. *Frequently Asked Questions about Huanglongbing (HLB; Citrus Greening) for Homeowners.* University of Florida IFAS Extension. https://edis.ifas.ufl.edu/pdffiles/PP/PP32600.pdf (2016).
21. Court, C. D., Hodges, A. W., Rahmani, M. & Spreen, T. H. Economic Contributions of the Florida Citrus Industry in 2015–16. University of Florida IFAS Extension. https://fred.ifas.ufl.edu/media/fredifasufledu/photos/economic-impact/Economic-Impacts-of-the-Florida-Citrus-Industry-2015-16.pdf (2017).
22. Butler, S. Rude awakening as price of coffee and orange juice shoots up 20%. *The Guardian.* https://www.theguardian.com/business/2016/sep/30/rude-awakening-as-price-of-coffee-and-orange-juice-shoots-up-20 (2016).
23. Coletta-Filho, H., Gonçalves, F., Amorim, L., de Souza, A. & Machado, M. Survey of Xylella fastidiosa and citrus variegated chlorosis in Sao Paulo State, Brazil. *J. Plant Pathol.* **95,** 493–498 (2013).
24. Ghini, R., Bettiol, W. & Hamada, E. Diseases in tropical and plantation crops as affected by climate changes: Current knowledge and perspectives. *Plant Pathol.* **60,** 122–132 (2011).
25. Neate, R. Feeling the squeeze: Florida faces worst orange harvest crisis in a century. *The Guardian.* https://www.theguardian.com/us-news/2016/oct/16/florida-oranges-juice-harvest-disease-hurricanes (2016).
26. Morgan, K. T., Zotarelli, L. & Dukes, M. D. Use of irrigation technologies for citrus trees in Florida. *HortTechnology* **20,** 74–81 (2010).
27. Parsons, L. R. & Morgan, K. T. *Management of Microsprinkler Systems for Florida Citrus.* University of Florida IFAS Extension. http://edis.ifas.ufl.edu/hs204 (2004).

28. Fares, A. Citrus irrigation scheduling. *Tree For. Sci. Biotechnol.* **3**, 12–21 (2009).
29. Boman, B., Smith, S. & Tullos, B. Control and Automation in Citrus Microirrigation Systems. *Document no. CH194* (Institute of Food and Agricultural Science, University of Florida Gainesville, FL, 2002).
30. Futch, S. H. Brazil citrus tours increase Floridians' knowledge. *Citrus Industry.* https://crec.ifas.ufl.edu/extension/trade_journals/2013/2013_January_ brazil.pdf (2013).
31. Brito, R. P. d., Miguel, P. L. d. S. & Pereira, S. C. F. Impacts of natural disasters in Brazilian supply chain: The case of São Paulo drought. http://biblioteca-digital.fgv.br/dspace/bitstream/handle/10438/17778/Impacts_of_ Natural_Disasters_in_Brazilian_Supply_Chain_The_Case_of_São_Paulo_ Drought.pdf?sequence=1 (2016).
32. Davies, F. S. & Maurer, M. A. Reclaimed wastewater for irrigation of citrus in Florida. *HortTechnology* **3**, 163–167 (1993).
33. FAO. Managing huanglongbing/citrus greening disease in the Caribbean. *FAO Subregional Office for the Caribbean Issue Brief* **4**. http://www.fao.org/3/a-ax739e.pdf (2013).
34. Betancourt, M. *et al.* First quantitative estimates of carbon retention by citrus groves under Cuba's conditions. *Am. J. Clim. Change* **3**, 130 (2014).
35. Nakano, V. E. *et al.* Evaluation of pesticide residues in oranges from São Paulo, Brazil. *Food Sci. Technol.* **36**, 40–48 (2016).

CHAPTER 3

Climate-Smart Bread

Abstract Globally we produce 700 million tonnes of wheat each year, providing one-fifth of all the calories and proteins we consume. An average loaf of bread has a carbon footprint of 1 kilogram, mainly as a result of emission on the farm. We also waste a huge amount: over 700,000 tonnes is thrown away each year in the UK—the equivalent of more than two million loaves a day and about one-third of all the bread we buy. Alongside reducing household waste, improved efficiency of nitrogen fertiliser use is a key way to cut emissions. Wheat is already facing impacts on yields from climate change, with heat waves, drought and disease being major risks in many areas in the coming decades. Access to disease-resistant varieties and use of improved soil management can both boost resilience and reduce emissions.

Keywords Wheat • Flour • Food waste • Carbon footprint • Heat stress • Fusarium • Mycotoxins • Disease resistance • Soil organic matter • Drainage • Precision agriculture

Since the dawn of agriculture, bread has been the yeast-heavy fuel for much of human civilisation. Most is now made using wheat flour, with a seemingly endless array of shapes, flour mixes and bakes available. Millions of working days begin with a slice or two of toast, hastily smeared with butter and bolted down with one hand on the car keys, umbrella or screaming toddler [1]. Here in Scotland, we eat our way through over 700,000 loaves each day [2], with most of these being the white pre-sliced favourite of sandwich and toast-making.

In our house, the misshapen rolls and part-blackened loaves on offer are usually courtesy of my own amateur baking efforts. These unpredictable experiments in bread-making rely on bags of strong flour in each weekly food shop. Though the bulk of bread flour sold here in the UK is home grown, up to one-fifth is imported from Canada due to the particular strengthening qualities their flour possesses. Globally though, it is China and the US that lead the way on wheat production, with Russia and India also being big players (Fig. 3.1).

Wheat is a true global staple. As well as giving us bread, it is the basis for myriad pastas, cakes, breakfast cereals and snacks. It's also a major food source for livestock. The 700 million tonnes produced worldwide each year [4, 5] provide one-fifth of all the calories and proteins we consume

Wheat production, 2014
Annual agricultural production of wheat, measured in tonnes per year.

Our World in Data

| No data | 0 t | 10 million t | 50 million t | 200 million t |
| | 5 million t | 25 million t | 100 million t | |

Fig. 3.1   Global wheat production in 2014 by country of origin (Source: Hannah Ritchie, Our World in Data) [3]. Available at: https://ourworldindata.org/grapher/wheat-production

[6], and there are now more than 200 million hectares of land dedicated to its production (the biggest area of any crop on the planet) [7].

Wheat-growing practice has seen huge leaps over the last 50 years. Pressure to wring bigger and bigger yields from the world's fields has meant burgeoning use of artificial fertilisers, high-yield varieties, pesticides and irrigation [8]. The initial results were stunning. As access to the powerful tools of this Green Revolution spread from developed nations to developing nations in the 1960s, so yields surged across Latin America, Asia and North Africa. The new high-yield wheat varieties had shorter stems, and so were less prone to falling over in high winds. They could also reach maturity faster and better convert added fertilisers into bumper crops.

This bonanza period for global wheat production could not be sustained. By the 1980s the big year-on-year increases in harvests were starting to falter, with problems of unequal access to fertilisers and new crop varieties being common. By the 1990s and early 2000s increasing numbers of farmers were finding that crop losses, due to problems like increasing soil salinity and waterlogging, were undermining the yield-boosting effects of big fertiliser, pesticide and irrigation inputs.

Gradually emerging as a fundamental threat to global wheat production over this period has been climate change [9]. Rising carbon dioxide concentrations in the atmosphere initially seemed a boon—the feast of nutrients supplied by artificial fertilisers together with big improvements in water management allowing the crops to take fuller benefit of this carbon dioxide enrichment effect. Yet as carbon dioxide concentrations have risen, so global temperature has also increased and precipitation patterns changed. Since 1980 rising temperatures are estimated to have sliced around 5 per cent off global wheat yields [10].

For most wheat production, disentangling the effects of rising temperature from changes in water availability and farming practices remains difficult.

Yet warning signs of wheat's sensitivity to severe weather, and the knock-on effects for global food security, were already clear in 1972 when a combination of low rainfall and record temperatures hit yields in Russia. That summer peak temperatures pushed past 30 degrees Celsius just as the Russian wheat crop was developing [11]. Harvests fell by more than a tenth, with the decision by the Russian government to then buy from the global wheat market pushing up world prices and sparking fears of food shortages in other nations.

Today, much of the world's wheat is grown in areas that are experiencing both rising temperatures and falling water availability—a potential double-risk as the water demand of wheat plants increases as temperatures rise. The higher temperatures also tend to speed up plant growth and so shorten its growing time [12]. For the key period when the plants would normally be developing their grains, this can mean much smaller wheat grains are produced and overall yields can be more than halved [4].

While gradual increases in temperature may actually enhance wheat yields in some places (at high latitudes where low temperatures currently restrict crop growth for instance), it is an increase in extreme events like heat waves and droughts that poses the biggest risk in many areas [5]. Recent decades have seen the High Plains farmers of Dakota and Montana reaping the benefits of a warming climate, overtaking Kansas as top wheat-producing area in the US. Yet in 2017 this same area was hit with one of the most severe droughts on record.

As the wheat-growing season began, so the rainfall petered out. By August rainfall across the region was down to half the normal level, with some places receiving less than a quarter of their expected rain [13]. Described as a flash drought because of its sudden onset and severity, the combination of thirsty plants, high temperatures and vanishing rainfall hit both wheat yield and quality hard [14, 15]. Total production in the region was down by around a quarter compared to the average, with over 200,000 hectares of North Dakota wheat fields never making it through to harvest. The speed and unprecedented nature of this drought in some areas made pre-emptive action by farmers near impossible. Droughts predictions are given each year, based on factors like the size of the spring snowpack and seasonal temperatures, but these forecast methods can still be side-swiped by flash droughts [16].

North Dakota has already seen the fastest increase in average temperatures anywhere in the contiguous US [17]. With further warming (the number of days over 100 degrees Fahrenheit is set to double by 2050) and even larger swings in water availability predicted, the High Plains farmers face a choice between investing in expensive irrigation systems or trying to build these new extremes into their management plans [14].

For our own great British loaf, wheat farmers in the UK have largely been spared such flash drought and high temperature threats. Instead, it is the challenge of too much water, at the wrong time and in the wrong place that most risks damaging yields and further inflating the climate cost of our daily bread.

The carbon footprint of bread depends on how and where the wheat is grown, its processing and production, and how we then choose to consume it [18]. Across its life cycle the average family loaf—including my own home-baked blobs—will clock up around one kilogram of emissions. Growing the wheat is the biggest part of this, at one-third of the total, and mostly arises from the nitrogen fertilisers used to boost yields. Consumer use of the bread—such as the energy used if we keep it in the fridge or toast it—then makes up another quarter of the emissions. The rest comes from processing, packaging, transport and other ingredients like salt, sugar and yeast. Although bread made with imported flour can have a slightly higher carbon footprint, transport emissions are only a small part of this difference, with the wheat-growing practices of the source farms being much more important.

For households, using all the bread that we buy, rather than wasting it, is the most powerful way for most of us to reduce its carbon footprint—in the appetite-killing world of food waste statistics bread is in a class of its own. In the UK over 700,000 tonnes are thrown away each year—the equivalent of more than two million loaves a day and about one-third of all the bread we buy. The climate cost is a global embarrassment at over half a million tonnes of greenhouse gas.

Almost all is classed as avoidable and is mainly due to the bread not being used in time [19]. Additional waste arises from personal tastes and that perennial sandwich leftover: crusts. Not over-buying, keeping it well wrapped or frozen [20], and using even those tantrum-inducing crusts can therefore avoid unnecessary land, fertiliser and pesticide use, and so play a major role in lowering bread's overall carbon footprint. In wheat fields around the world a similar battle to prevent loss and waste is one that has been fought by generations of farmers.

* * *

Most UK wheat—some two million hectares providing a harvest worth over £2 billion a year—is grown as so-called winter wheat. The seeds are sown in the autumn and then harvested in July to September of the following year. As climate change intensifies the UK is set to experience drier, hotter summers alongside warmer, wetter winters. Since the 1960s, average summer temperatures have risen by around 1 degree Celsius [21] and rainfall has dropped by a tenth in many regions [22]. For the bulk of wheat farmers this trend has helped lengthen the growing season and, along with new technology and wheat varieties, has boosted yields [23]. However, winter precipitation has increased markedly too, with all regions getting wetter and some areas of Scotland now seeing 60 per cent more rain and snow.

On heavy clay-rich soils, persistent rains and waterlogging can induce low oxygen conditions that damage plant roots, while precious nutrients and soil organic matter are lost to field drains via run-off and leaching. A really heavy downpour will also bend wheat stems, pushing the drenched plants into the dark, moist and windless under-storey of the wheat crop where pests and fungal diseases can thrive [24].

In 2012 Britain endured one of its wettest years on record. After a warm and dry start to the year, accompanied by initial concerns over drought, England and Wales then saw their wettest April and June since 1766 and the wettest summer season overall since 1912 [25]. By that autumn, soils were saturated and the rain just kept on coming [26]. Hundreds of homes and businesses were flooded, transport links were cut, and costs spiralled to near £600 million [27]. Many farms were badly hit. Honey production was more than halved, apple growers faced their worst harvest in 15 years [28], and potato farmers their worst in over three decades [29].

In comparison wheat got off lightly—the total harvest in the UK was down 13 per cent on the previous year [30]. Farmers then faced the choice between not planting winter wheat at all or trying to sow their sodden fields with the risk of bogged-down machinery, damaged soil structure and poor harvests in the following year too. Many opted for the former, with 400,000 hectares of land normally planted with wheat either used for growing different crops or abandoned to the winter rains [31].

The 2013 wheat harvest was inevitably a poor one. Some farmers who had initially braved the muddy fields found wheat growth so poor that they grubbed up the struggling crop and planted with barley instead [32]. For the first time in over a decade the UK became a net importer of wheat as total yields slipped a further 8 per cent, compounding the losses of 2012 [31]. Grain prices were pushed to record highs, with the price of everything from bread and breakfast cereals to beer and beef being affected.

The waterlogged and abandoned wheat fields of 2012 and 2013 were an all too obvious sign of severe weather impacts, but hidden within most swaying stands of outwardly healthy wheat a host of additional threats lurk that may become super-charged by climate change.

Wheat has many enemies. Globally, weeds are one of its biggest adversaries, hitting yields by crowding out seedlings and capturing the water and nutrients intended for the crop. The swathes of closely packed wheat plants now grown in much of the world also represent a rich target for any pests and diseases able to invade them. Worldwide, bacteria, viruses and fungal diseases, already have major impacts on yields. Between 2001 and 2003 fungal and bacterial disease were responsible for an estimated 10 per cent of global wheat losses [33].

Through changes in rainfall, temperature and plant growth, climate change will alter the spread and impact of such diseases in many areas [7]. In the moist air and soils of the UK, fungal diseases—ranging from 'rusts' and 'blights' to 'smuts' and 'bunts'—now pose a major threat. Public enemy number one for many British wheat farmers are fungi belonging to the *Fusarium* family. Given warm and wet conditions these fungi can invade the wheat plants just as they are starting to flower. The fungal blight appears as bleaching of the ears of wheat, along with clusters of pink or orange fungal spores [34]. In bad years the spread of this Fusarium Ear Blight in the UK can reach epidemic proportions, with eight out of ten wheat crops sampled in 2007 showing signs of infection. As a result there can be big losses in yield and quality, but the most damaging impact is often a more cryptic and highly toxic one. The invading fungi can produce toxins (called mycotoxins) that, if present in the wheat grains, are dangerous to any animals or humans that then consume them [35]. In livestock, consumption of heavily contaminated grain can lead to loss of appetite, weight loss and vomiting. In humans its effects may include dizziness, abdominal pain and fever.

How exactly climate change will affect fungi and other wheat pests and diseases around the world remains uncertain. In the UK, modelling studies suggest an increase in the intensity of *Fusarium* epidemics, especially in the south of England, through to 2050 [34]. Likewise, an increase in *Fusarium* infections in some parts of Brazil and Uruguay is expected, while some other fungal diseases (such as the charmingly named Karnal Bunt) may actually decrease in the Punjab region of India due to changing temperature and humidity [36]. A future atmosphere enriched with more carbon dioxide adds to the uncertainty—though wheat plants may grow faster, their many pests and diseases could also benefit [37].

What *is* certain is that a changing climate will indirectly affect wheat yields through its impacts on weeds, pests and diseases, as well as through the more direct effects of changing temperature and rainfall on plant growth. On its trajectory of warmer, wetter winters, and increasing risks of extreme weather, pest and disease impacts, UK wheat production is set to experience the same rollercoaster ride of peaks and troughs being faced around the world. Where these troughs are global, and wheat prices spike, they could pose a very real danger to the food security of millions.

*   *   *

A climate-smart response for wheat growers, wherever they are in the world, will need to try and take account of both the direct and indirect risks from climate change. Dealing with heavy rainfall and saturated soils has become second nature for many British farmers, with a proven adaptation response being to avoid planting wheat altogether in very wet years. In some areas with high rainfall or heavy, clay-rich soils, installing or improving land drainage can help to prevent waterlogging and ensure that planting, growth and harvest can still go ahead. However, drainage can be a costly solution—it usually involves permeable pipes being laid at regular intervals under the soil and so requires specialised equipment, and good maintenance. In some cases these soil drains also increase the risks of nutrients being carried away from the fields along with the drainage water [38].

Careful management of soils and their structure is often a very effective way to reduce the need for extra drainage and to boost the resilience of the fields to extreme rainfall, as well as to many drought, pest and disease risks. Reducing compaction of the soils for instance, by avoiding heavy machinery use when they are saturated, helps keep the air spaces within the soil open and so maintains their permeability. Likewise, increasing the organic matter in soils [39]—such as by incorporating manures and reducing the amounts and depth of tillage—can improve soil structure and its drainage properties [40]. Such deliberate management of carbon in soils is a climate-smart food solution that can truly pull off the triple-win trick of increasing climate change resilience, boosting yields and enhancing carbon stocks [39].

For those wheat farmers able to afford them, a suite of new technologies may play an important role in maximising yields and dealing with weather extremes. Accessing waterlogged fields without causing severe damage to soil structure has been made easier with the development of lighter tractors and caterpillar tracks that spread weight better. The expansion of precision agriculture—whereby nutrients and pesticides can be delivered more precisely to where they are needed—now includes satellite data, soil moisture sensors, and even remotely controlled drones that are used to survey crops from the air and identify waterlogged, diseased or under-fertilised areas that need attention [41].

Less eye-catching, but arguably more important, is the widespread practice of soil monitoring, farm nutrient budgeting, and the use of long range weather forecasting to help inform planting and field management decisions [42, 43]. An increasing number of 'decision-support' tools are available to wheat farmers, ranging from free-to-use online applications specifically aimed at reducing greenhouse gas emissions from agriculture [44], to pay-for platforms that claim to provide seasonal forecast and weather data at the level of individual fields up to 3 months in advance [45].

For major wheat pests and diseases like *Fusarium* blight there are similar online tools and advice designed to support growers and assess the level of risk based on location, time of year and weather conditions [46]. As well as helping to give early warning of potential disease outbreaks, these sources of information include guidance on how best to harvest, store and process the wheat so as to minimise infection and toxins, and therefore cut overall losses. As global trade in wheat increases—including years of big wheat imports to the UK—the risk of disease spread is likely to grow and the need for early warning and robust quarantine systems becomes even more important.

Development of new types of disease-resistant wheat is a potentially very powerful way to increase resilience [47], with the search for wheat plants that are both higher yielding and better adapted for future climate change having become a global undertaking [48]. A big question in the development and use of climate-resilient wheat is that of exactly what the main climate change risks will be in any given location, and so which particular traits should be emphasised.

Developing plants that are more resistant to fungal infection might appear a good option for the UK, yet increased resistance to one threat can come at the cost of reduced defence against others [23, 49]. Understanding such trade-offs and preparing for future climate risks is a focus of the research that is currently underway. In some parts of Africa for instance, it is estimated to take over ten years for controls such as crop resistance to a new disease to be put in place [50]. If the risks of drought, flood or disease under a changing climate can be identified early then their impacts could be drastically reduced.

As with any type of food, the most effective strategies for climate-smart wheat production will ultimately be those that are specific to the local context, that are cost-effective and that are readily accessible to farmers. The high level of expertise common to British wheat farmers, coupled with good access to technology, advice and cutting-edge research, means they are well placed to adapt to future climate change. There could be opportunities to expand wheat into new areas of the UK as growing seasons lengthen, while future summers may bring more of the heat and drought stress challenges already faced by farmers in the High Plains of the US [23].

Nationwide, there are still gaps in the provision of climate extension services (advice and support for farmers), and more locally specific assessments of climate risks and opportunities are definitely needed. However, these shortfalls are dwarfed by those faced by wheat farmers in most of the world. As the human population continues to rise and wheat demand spirals upward, climate change impacts pose a major risk for sustaining current harvests, let alone increasing them, in many nations [51]. Each degree centigrade of climate warming is predicted to slice another 6 per cent from global wheat production [23]. With one-fifth of our calories coming from wheat and its vulnerability to climate change being so high [52], making our daily bread more climate-smart is arguably one of the most important challenges facing agriculture in the twenty-first century.

REFERENCES

1. fabflour.co.uk. *Facts about Bread.* http://fabflour.co.uk/fab-bread/facts-about-bread/ (2018).
2. nabim.org.uk. *Flour & Bread Consumption.* http://www.nabim.org.uk/flour-and-bread-consumption (2018).
3. Ritchie, H. Global wheat production, 2014. *Ourworldindata.org.* https://ourworldindata.org/grapher/wheat-production (2018).
4. Asseng, S., Foster, I., & Turner, N. C. The impact of temperature variability on wheat yields. *Glob. Change Biol.* **17**, 997–1012 (2011).
5. Zampieri, M., Ceglar, A., Dentener, F. & Toreti, A. Wheat yield loss attributable to heat waves, drought and water excess at the global, national and sub-national scales. *Environ. Res. Lett.* **12**, 064008 (2017).
6. Shiferaw, B. *et al.* Crops that feed the world 10. Past successes and future challenges to the role played by wheat in global food security. *Food Secur.* **5**, 291–317 (2013).

7. Juroszek, P. & von Tiedemann, A. Climate change and potential future risks through wheat diseases: A review. *Eur. J. Plant Pathol.* **136**, 21–33 (2013).

8. Pingali, P. L. Green revolution: Impacts, limits, and the path ahead. *Proc. Natl. Acad. Sci.* **109**, 12302–12308 (2012).

9. Rosenzweig, C. & Parry, M. L. Potential impact of climate change on world food supply. *Nature* **367**, 133–138 (1994).

10. Lobell, D. B., Schlenker, W. & Costa-Roberts, J. Climate trends and global crop production since 1980. *Science*, 1204531 (2011).

11. Battisti, D. S. & Naylor, R. L. Historical warnings of future food insecurity with unprecedented seasonal heat. *Science* **323**, 240–244 (2009).

12. Iqbal, M. *et al.* Impacts of heat stress on wheat: A critical review. *Adv. Crop Sci. Technol.* **5**, 1–9 (2017).

13. Holthaus, E. 'Flash drought' could devastate half the High Plains wheat harvest. *grist.org.* https://grist.org/food/flash-drought-could-devastate-half-the-high-plains-wheat-harvest/ (2017).

14. McLaughlin, K. The unprecedented drought that's crippling Montana and North Dakota. *The Guardian.* https://www.theguardian.com/environment/2017/sep/07/flash-drought-north-dakota-montana-wildfires (2017).

15. Rooney, D. US winter wheat ratings slump in key growing states. *AHDB Cereals & Oilseeds.* https://cereals.ahdb.org.uk/markets/market-news/2018/january/30/grain-market-daily-300118.aspx (2018).

16. Otkin, J. Hit hard by drought. *toolkit.climate.gov.* https://toolkit.climate.gov/case-studies/early-warning-information-increases-options-drought-mitigation (2017).

17. USEPA. *Climate Impacts in the Great Plains.* https://archive.epa.gov/epa/climate-impacts/climate-impacts-great-plains.html (2017).

18. Espinoza-Orias, N., Stichnothe, H. & Azapagic, A. The carbon footprint of bread. *Int. J. Life Cycle Assess.* **16**, 351–365 (2011).

19. WRAP. Household food and drink waste in the United Kingdom 2012. *Waste and Resource Action Programme.* http://www.wrap.org.uk/sites/files/wrap/hhfdw-2012-main.pdf.pdf (2013).

20. Taylor, A.-L. Why is bread Britain's most wasted food? *BBC Online.* http://www.bbc.co.uk/news/magazine-17353707 (2012).

21. Carbonbrief. Mapped: How every part of the world has warmed—And could continue to warm. *Carbonbrief.org.* https://www.carbonbrief.org/mapped-how-every-part-of-the-world-has-warmed-and-could-continue-to-warm (2018).

22. Jenkins, G. J., Perry, M. C. & Prior, M. J. *The Climate of the UK and Recent Trends.* UK Met Office. http://www.ukcip.org.uk/wp-content/PDFs/UKCP09_Trends.pdf (2008).

23. Brown, I. *et al.* UK Climate Change Risk Assessment Evidence Report: Chapter 3, natural environment and natural assets. *Report prepared for the*

Adaptation Sub-Committee of the Committee on Climate Change, London. (2016).

24. Mäkinen, H. *et al.* Sensitivity of European wheat to extreme weather. *Field Crop. Res.* **222**, 209–217 (2018).

25. MetOffice. *UK Weather—Annual 2012.* https://www.metoffice.gov.uk/climate/uk/summaries/2012/annual (2013).

26. MetOffice. *November 2012 Flooding.* https://www.metoffice.gov.uk/learning/learn-about-the-weather/weather-phenomena/case-studies/november-2012-flooding (2015).

27. UKEA. *Floods Cost UK Economy Nearly £600 Million.* http://webarchive.nationalarchives.gov.uk/20140223163833/http://www.environment-agency.gov.uk/news/150962.aspx?page=2&month=11&year=2013 (2013).

28. Morris, S. Apple growers face grim harvest with worst yield for 15 years. *The Guardian.* https://www.theguardian.com/lifeandstyle/2012/sep/16/apple-growers-worst-yield-15-years (2012).

29. fwi.co.uk. *2012 Potato Crop Lowest Since 1976.* http://www.fwi.co.uk/arable/2012-potato-crop-lowest-since-1976.htm (2012).

30. DEFRA. *Farming Statistics: Final Crop Areas, Yields, Livestock Populations and Agricultural Workforce at 1 June 2012, United Kingdom.* Department for Environment, Food & Rural Affairs, UK. http://webarchive.nationalarchives.gov.uk/20130125173510/http://www.defra.gov.uk/statistics/files/defra-stats-foodfarm-landuselivestock-farmingstats-june-statsrelease-june12fina-luk-121220.pdf (2012).

31. DEFRA. *Farming Statistics: Provisional Crop Areas, Yields and Livestock Populations at June 2013, United Kingdom.* Department for Environment, Food & Rural Affairs, UK. https://www.gov.uk/government/uploads/system/uploads/attachment_data/file/251222/structure-jun2013prov-UK-17oct13a.pdf (2013).

32. Vidal, J. Farmers fail to feed UK after extreme weather hits wheat crop. *The Guardian.* https://www.theguardian.com/environment/2013/jun/12/farmers-fail-weather-wheat-crop (2013).

33. Oerke, E.-C. Crop losses to pests. *J. Agri. Sci.* **144**, 31–43 (2006).

34. Madgwick, J. W. *et al.* Impacts of climate change on wheat anthesis and fusarium ear blight in the UK. *Eur. J. Plant Pathol.* **130**, 117–131 (2011).

35. FAO. *Mycotoxins in Grain.* Food and Agriculture Organization of the United Nations. http://www.fao.org/wairdocs/x5008e/x5008e01.htm (1997).

36. Kaur, S., Dhaliwal, L. & Kaur, P. Impact of climate change on wheat disease scenario in Punjab. *J. Res.* **45**, 161–170 (2008).

37. Bencze, S., Vida, G., Balla, K., Varga-László, E. & Veisz, O. Response of wheat fungal diseases to elevated atmospheric CO2 level. *Cereal Res. Commun.* **41**, 409–419 (2013).

38. AHDB. *Field Drainage Guide.* Agriculture and Horticulture Development Board (AHDB). https://cereals.ahdb.org.uk/media/725158/g68-ahdb-field-drainage-guide.pdf (2015).

39. Paustian, K. *et al.* Climate-smart soils. *Nature* **532**, 49 (2016).

40. POST. Securing UK soil health. *UK Parliament POSTNOTE Number 502.* researchbriefings.files.parliament.uk/documents/POST-PN-0502/POST-PN-0502.pdf (2015).

41. Vidal, J. Hi-tech agriculture is freeing the farmer from his fields. *The Guardian.* https://www.theguardian.com/environment/2015/oct/20/hi-tech-agri-culture-is-freeing-farmer-from-his-fields (2015).

42. AHDB. *Monitoring.* https://cereals.ahdb.org.uk/monitoring.aspx (2018).

43. SRUC. *Nutrient Budgeting.* https://www.sruc.ac.uk/info/120605/soil_and_nutrients/1551/nutrient_budgeting (2018).

44. CFA. *The Cool Farm Tool.* https://coolfarmtool.org/coolfarmtool/ (2016).

45. weatherlogistics.com. *Seasonal Weather Predictions.* https://www.weatherlo-gistics.com (2018).

46. AHDB. *Fusarium Infection Risk Archive.* AHDB Cereals & Oilseeds. https://cereals.ahdb.org.uk/monitoring/fusarium/fusarium-infection-risk-archive.aspx (2018).

47. Nicholson, P., Bayles, R. A. & Jennings, P. *Understanding the Basis of Resistance to Fusarium Head Blight in UK Winter Wheat (REFAM)* (Home-Grown Cereals Authority, 2008).

48. Mondal, S. *et al.* Harnessing diversity in wheat to enhance grain yield, climate resilience, disease and insect pest resistance and nutrition through conventional and modern breeding approaches. *Front. Plant Sci.* **7**, 991 (2016).

49. Wulff, B. B. & Moscou, M. J. Strategies for transferring resistance into wheat: From wide crosses to GM cassettes. *Front. Plant Sci.* **5**, 692 (2014).

50. Newbery, F., Qi, A. & Fitt, B. D. Modelling impacts of climate change on arable crop diseases: Progress, challenges and applications. *Curr. Opin. Plant Biol.* **32**, 101–109 (2016).

51. Lobell, D. B. *et al.* Prioritizing climate change adaptation needs for food security in 2030. *Science* **319**, 607–610 (2008).

52. Lobell, D. B. & Field, C. B. Global scale climate–crop yield relationships and the impacts of recent warming. *Environ. Res. Lett.* **2**, 014002 (2007).

CHAPTER 4

Climate-Smart Tea

Abstract Tea is second only to water as the world's most popular drink. Total production has grown to around 5 million tonnes a year, with the industry as a whole valued at $20 billion. Its carbon footprint totals 25 grams of emissions per comforting cup, mainly the energy used to boil water. In the UK almost 60,000 tonnes is wasted each year simply because too much was made or it went cold before we could drink it. Tea plants face climate risks from flood, drought and heat, as well as from increases in disease and pest attack, such as the already highly damaging tea mosquito bug. The use of agroforestry, soil management and carefully planned drainage can all help increase resilience to severe weather impacts. New tea varieties also offer the chance to increase disease resistance. Moving cultivation to new areas is likely to be the only option in some areas as the climate envelop for tea shifts further in the coming decades

Keywords Assam • Agroforestry • Heat stress • Carbon footprint • Shade trees • Drainage • Soil management • Tea mosquito bug • Cover crops • Mulch • Rainfall capture

The crumbled leaves that infuse the daily cuppa of billions are still largely handpicked just as they've been for centuries. Harvesting machines might be faster, but these also tend to mangle the leaves and can damage the precious bushes. Instead, just the top bud and two leaves on new shoots are carefully plucked every week or so in early spring, and then again in summer. This regular plucking keeps the bushes in a perpetual state of neat new growth and ensures they are always at a convenient table-top height for harvesting [1]. Tea bushes are a long-term investment, with many being over 50 years old and so needing generations of care and attention to ensure they produce the very best quality leaves [2].

Like much of our daily food and drink, tea tends to clock up some very long distances between plantation and shopping trolley. Though China is now the world's biggest producer (Fig. 4.1), much of it is still produced by the tea-growing giant that is India, with regions like Darjeeling and Assam being world-famous for the quantity and quality of their tea.

Our own neat box of aromatic tea bags contains a blend grown in Assam—a high plateau in the north of India that is responsible for over half of all Indian production [4]. In terms of its carbon footprint, the growing and processing (sorting, drying and the oxidation that turns leaves from green to black) are the main sources of emissions before our

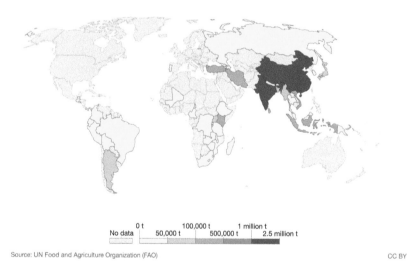

Tea production, 2014
Annual agricultural production of tea, measured in tonnes per year.

Source: UN Food and Agriculture Organization (FAO)

Fig. 4.1 Global tea production in 2014 by country of origin (Source: Hannah Ritchie, Our World in Data) [3]. Available at: https://ourworldindata.org/grapher/tea-production

tea reaches us [5]. Despite long travel distances, transport emissions are relatively low with most being exported by sea using large shipping containers. Once in our cupboards and caddies though, making the tea with boiling water becomes the biggest component of the overall carbon footprint, weighing in at around two-thirds of the total 25 grams of emissions per comforting cup (equivalent to 12 kilograms per kilogram of dry tea) [6].

One helpful trend that should cut the carbon footprint of each cup of tea is the move towards more renewable energy. As more comes on line, so the electricity that powers our many millions of kettles is becoming lower carbon. Even more important at the household level is cutting waste. First, the boiling of too much water leads to wasted energy use and so emissions—boiling half a litre of water, instead of the quarter litre actually required, will balloon the carbon dioxide footprint of an average cup of Kenyan tea from a trim 25 grams to a far-weightier 42 grams [6] (around three-quarters of us still boil more water than we need [7]).

Then comes the covert climate penalty incurred by wasting the tea that has been made. Each year in the UK, households generate around half a million tonnes of waste from the tea they make. Most of this is unavoidable, such as used tea bags and leaves destined for the compost bin, but almost 60,000 tonnes is wasted simply because too much was made or it went cold before we could drink it [8]. The emissions avoided by making only what is needed and reheating those forgotten lukewarm cuppas really add up—with a life-cycle carbon footprint that can top 10 kilograms for each kilogram of tea, current wastage racks up hundreds of thousands of tonnes of unnecessary greenhouse gas every year.

Our insatiable thirst for the liquid gold that is tea dates back millennia. Threats to the world's precious supplies have been enough to spark protests, blockades and even wars in the past [9]. In the eighteenth century, Britain's own efforts to secure imports by fixing the market resulted in American colonists showing their anger by throwing a shipment of tea into Boston harbour. This Boston Tea Party signalled the demise of British rule in America.

Today, tea is second only to water as the world's most popular drink. Total production has grown to around five million tonnes a year [10] with the industry as a whole valued at $20 billion [11]. The days of a British empire wielding its power to manipulate global tea markets may be long gone, but in the lush fields of northern India climate change is laying a heavy hand on the supplies of millions.

Assam is well known for the full-bodied flavour of its tea, with the neatly clipped swathes of bushes that cover its rich floodplains usually lux-uriating in a tropical mix of monsoon rains and temperatures in the high-20s Celsius. In recent years though, things have become more unpredictable, and much more dangerous. In the summer of 2013 tem-peratures in Guwahati—Assam's economic centre—soared into the high-30s Celsius, breaking previous records, forcing schools to shut and causing several deaths from heat stress [12]. The following year, more high tem-peratures combined with months of little or no rainfall to trigger drought warnings across the state [13]. Then, in 2016, torrential rains caused land-slides and floods that took the lives of more than 60 people [14]. As well as the immediate risk to life and limb, these waves of extreme weather events also signalled dangerous times for the livelihoods of Assam's esti-mated 1.2 million tea workers [2].

*   *   *

The tea gardens of Assam thrive on the plentiful rains that sweep over them each summer. With more extreme wet spells though [15], a triple threat emerges. First, heavy downpours erode soils and strip away precious nutrients. Bushes in low lying and poorly drained soils may find their roots become waterlogged. If this goes on too long then the roots begin to rot and survival of the whole plant is put at risk [16]. Tea bushes that escape waterlogging then have to run the gauntlet of the pests and diseases that can thrive in the wetter conditions.

Tea plays host to an army of pests and diseases that between them can attack every part of the plant. Diseases like Violet Root Rot are a common risk in heavy wet soils, while the airborne Blister Blight fungus attacks young shoots and can spread rapidly in moist, foggy conditions [1]. Tea plant pests are broadly divided into those that chew on the plants and those that suck on them. Of most concern to tea farmers across India and globally is one of the suckers: the Tea Mosquito bug [17].

Both the nymphs and the adults of the Tea Mosquito draw sap from newly emerged buds and shoots. The affected areas first develop irregular brown spots, the leaves then begin to dry and curl up. Eventually the mul-tiple sucking injuries turn the new-growth leaves black and unusable. In bad cases the growth of the whole plant is stunted. Even mild infestations can cause lasting damage, as when the eggs of the Tea Mosquito bug are laid on the plants' stems they cause them to crack and dieback.

These bugs were first recorded in Java in the nineteenth century, then in India in 1968, and have since become a menace to tea growers all over the world [17]. They have been blamed for large yield losses in Africa, and have destroyed entire tea harvests in some parts of Asia. In India, eight out of ten tea-growing areas are now infected, causing up to half of the annual crop to be lost.

As temperature and rainfall patterns change, the spread and impact of this sucking pest are set to change too. Higher rainfall, humidity and temperature have all been linked to faster bug population growth and an increase in attacks on the leaves of tea. Likewise, more rainy days in North Eastern India seem to allow the bugs to extend their attacks much later into the growing season. The powerful (and now banned) pesticide DDT was apparently effective at controlling the initial outbreaks that hit India in the late twentieth century. Since then a variety of other insecticides have been tried, but the bugs have become resistant to many and their march through the tea gardens of Assam has continued.

After extreme wet spells, a real sting in the tail for Assam's tea farmers comes when they are finally able to access their fields and harvest any surviving crop. When tea plants grow in very wet conditions key chemicals in their leaves are effectively diluted down [18]. These compounds define tea quality and help give Assam's strong breakfast tea its distinctive flavour. So, on top of losing some bushes to root rot, others to Tea Mosquito bug attacks, and having a poor harvest overall, the tea itself is often of low quality and hard to sell.

It's the increasingly unpredictable nature of rainfall that is posing the most immediate concerns for Assam's tea growers. While during one part of the growing season they must cope with the deluges and waterlogging problems of too much rain, at other times it might be severe drought that threatens the tea gardens. Such impacts of climate change are likely to be felt everywhere through major price hikes. If the kind of losses predicted for major tea-growing nations, such as India and Sri Lanka, are realised, then the cost of tea is set to increase by more than a quarter over the next decade [19].

Like all plants, tea has an optimal temperature range. This can vary somewhat between different varieties, but most thrive in the well-watered warmth of low to high-20s Celsius common to big tea-growing regions such as Assam. Go too far above or below this range and growth will slow or stop altogether.

Over the past century in Assam temperatures have been increasing [4, 20]. As the region warms further through the twenty-first century so the risk increases that its tea-friendly climate envelope will begin to shift and disintegrate. At the same time as average monthly temperatures in the summer have crept up, so the tea yield has waned. For every degree they go above 28 degrees Celsius the yield is cut by around 4 per cent [2]. Projected average temperatures by the middle of this century are some 2 degrees Celsius higher than today [21] and so, without adaptation, tea yields in Assam could keep on falling.

Whether extreme heat waves—with temperatures soaring into the high-30s Celsius—will cause even bigger impacts remains unclear. So far, most tea plants in Assam seem able to cope with short periods of such extreme heat [2], though for the region's multitude of tea pickers such heat waves bring a real risk of heat stress and illness [22].

For the plants it's a combination of higher temperatures with longer drought periods that can pose a particular risk. Under such conditions they conserve their diminishing water supplies by shutting down photosynthesis, and so growth. Hot conditions mean that soils dry out much more quickly and, with longer periods with no rain, the whole year's tea harvest can be knocked back [2]. The rains in the south bank area of Assam have decreased over the last 30 years [4], with a series of dry winters and springs meaning that the tea gardens are often already short on water before the growing season begins. Total rainfall in Assam by 2050 is not expected to change all that much, but its timing and the potential for more intense periods of both drought and flood pose a very real risk to its future as a tea-growing nirvana [4].

* * *

With many tea growers around the world facing yields being halved or worse as extreme weather impacts intensify, efforts to boost resilience have taken on renewed urgency. Much of the progress on making tea production climate-smart has been focused on water, and how best to deal with too much and too little.

To tackle heavy rainfall, and the waterlogging of plant roots that can then occur, soil drainage can be a very effective, with optimal drainage estimated to boost yields by up to 50 per cent [4]. Actually achieving such optimal drainage is difficult. Too much, and the tea plants might well stay safer in heavy monsoon rains, but will then be especially vulnerable to drought stress in the dry season. The undulating topography and patchwork of tea gardens in Assam makes getting drainage just right for every patch a complex task. Changing drainage in one place may increase flood and waterlogging risks in another, while widespread drainage can mean nutrient and pesticide leaching during rainstorms is made even worse.

As more intense downpours risk more soil erosion, adaptations that protect the soils can help protect the whole tea garden [4]. Mulching around plants and incorporation of composts help to improve soil structure and boost water-holding capacity—a boon for nutrient retention and reduced waterlogging, as well as drought resilience. The mulches and composts come from everything from old tea waste and weeds, to worm casts and tea clippings.

In some areas, so-called cover crops serve the dual purpose of soil protector and compost source. These short-lived crops are often grown alongside young tea plants to supress weeds. Many are legumes and so are also able to pull off the microbial root magic that is fixing nitrogen from the air. This then reduces the need for artificial fertilisers too. Once large enough, the cover crops become a nutrient-rich green manure that can then be incorporated back into soils.

Trees can also make an excellent growing companion to tea bushes. Planted in amongst the tea plants or along the margins of the tea gardens as shelter belts, the trees help to stabilise soils and protect them from the scouring effects of any heavy rainfall. Such deliberate tree planting in agriculture—agroforestry—is a popular climate-smart practice for many crops around the world. It not only has the potential to buffer crops and soil from heavy downpours, but also gives added protection from extreme drought and heat, and can lock-up carbon dioxide to boot.

In Assam, trees rising up above the neatly clipped understory of tea are a common sight. The trees are carefully spaced and nurtured so as to give the best balance of light, temperature and water possible—too much shade risks slowing tea growth and encouraging more attack from pests like the Tea Mosquito bug [1].

During the dry season the trees and their deep roots help to conserve soil moisture and protect the tea plants from extreme heat and scorching. The shade they cast improves the overall efficiency of tea growth, with their diffuse canopies also giving more protection from high winds and hailstorms. As a bonus, the steady fall of shade-tree twigs and leaves helps to boost the structure and organic matter content of the soils.

Shade trees can improve carbon storage too. The total amount of carbon locked up in the living biomass of these areas is estimated at over 50 tonnes per hectare [23]. Of this, the shade trees make up around 70 per cent of the stocks and, together with the tea bushes that surround them, sequester around five tonnes of carbon per hectare each year.

In the most severe hot and dry spells that are likely to beset Assam in coming years, shade trees and mulching may still not be enough. Here, the harvesting and reuse of rainwater can help to give tea growers that extra bit of protection. By creating reservoirs and ponds, farmers can capture more of the heavy monsoon downpours and then later use this same water to irrigate their plants and soils—applying any collected water in the early evening keeps the amounts lost to evaporation to a minimum and helps to bump up soil moisture levels when the plants are at their most thirsty [11].

As with any crop, relying on the success of just one type of tea growing in one particular place often increases the risk of failure of the whole harvest. Planting other crops with different harvest times and different vulnerabilities to drought, pests or floods can ensure that, if one harvest fails, there's at least another to deliver some income. In Assam, this kind of diversification is usually in the form of planting spice crops like black pepper alongside the tea gardens. Some of the more enterprising farmers have even taken to using their rainwater collection ponds to grow fish. The mud-loving Grass Carp is a favourite.

These carp are native to Asia and are well adapted for the weed-filled warmth of Assam's drainage ponds and reservoirs. They feed on submerged vegetation, and with good conditions, young fish can quickly grow to a harvestable size. These fishy grazing machines can also help to prevent drainage ditches and shallow ponds becoming clogged with vegetation—one trial release of grass carp to weed-choked ponds in India reportedly cleared the afflicted ponds within a month [24]. Its impressive ability to hoover up plant material has made the grass carp a popular choice for fish-livestock farming. Here, manure from ducks, pigs or other livestock is deliberately added to ponds to accelerate plant growth, which in turn boosts the growth of the fish and yields a useful extra income [25].

A wider mix of what is grown on tea plantations (including in the ponds) can therefore give farmers more resilience as they face a future of more frequent and intense extreme weather events. To tackle the increasing threat of pest and disease attacks though, such as from the Tea Mosquito bug, diversification of the tea plants themselves can be a powerful weapon in the climate-smart armoury.

We consumers are increasingly demanding tea that is free from all pesticide and herbicide residues, so tea growers are having to try and walk the tightrope between maintaining tea quality and keeping the tide of pests, diseases and weeds at bay. Tea plant clones help them to do this by providing plants selected for specific traits. The very first tea clone was developed and used in India in 1949 (before then all tea in the world was grown from seed). The clones are grown from selected cuttings and so allow characteristics such as resistance to a certain disease or greater drought tolerance to be brought out. As more and more clones have been developed, so the array of tea plants available to give resilience in different locations and conditions has expanded [4]. By definition, clones are identical, so tea growers tend to use them within a wider mix of tea plant types. This reduces the risk of one severe weather event or disease wiping out the whole crop—the new clone plants might all be good at giving resistance to a certain disease, but that's little comfort if they are then all destroyed by drought [26].

Where pesticides do still have to be used, the amounts can be limited by managing the tea gardens to keep pest numbers low and pre-empt their attacks. By more frequent plucking of leaves, more of the crop can be harvested before the pests get to the new shoots. Frequent plucking can also be directly targeted at the pests—removing the bugs from the plants before their numbers can increase to damaging levels. Using the natural enemies of tea pests has also become a successful ploy for many growers. These predators, such as some types of wasp and praying mantis, provide a natural control on pest numbers. They can be deliberately released to mop up infestations or can be encouraged by ensuring that any pesticides used target only the pests and leave these natural predators safe to do their work [1].

As climate change intensifies in Assam and around the world, so tea growers will need to make more and more use of a range of climate-smart practices. For some, the vagaries of tea prices combined with severe harvest losses will mean a move away from tea production altogether. For others, a change in location—migrating their tea gardens in line with the changing climate envelope—may be the only long-term solution.

Universal access to the various climate-smart techniques and management practices, from high-yielding tea clones and irrigation to diversification and soil improvement, remains a major barrier in many areas. Still, real progress is being made. Availability and use of improved tea clones is expanding, with real hope that the key to longer-term climate resilience may lie in the very DNA of the tea plants themselves.

In 2017 a team of Chinese researchers managed to sequence the entire tea genome [10]—an impressive feat given the tea genome is an especially large one made up of around 37,000 different genes. The work helped identify exactly where traits like drought, heat and pest resistance (as well as flavours) are derived from. For tea breeders, this genetic blueprint has huge potential to allow more climate change-resilient tea plants to be selected. It also opens up opportunities for targeted genetic modification to enhance those particular plant traits best suited for a particular region and its projected climate [27]. In Kenya, a new variety of purple tea has already been developed that is especially rich in antioxidants. These make it very valuable for use in health products and food preservation, and also more resistant to environmental stresses, pests and diseases [4].

Improving the links between tea growers, policy makers and researchers, to share best practice and jointly develop solutions, has been a focus of recent initiatives in Assam and elsewhere. By engaging directly with farmers and community leaders, training and field demonstrations can mainstream novel techniques and approaches. New market opportunities, such as generating income from carbon credits, certification and rebranding, could also be unlocked. Government involvement can provide climate-shock safety nets to tea farmers through guaranteed crop insurance and sustainable tea prices—while more investment in research will accelerate development and use of new tea varieties, climate change is not going to wait for us [4].

Globally, our most precious of hot drinks faces an uncertain future. Viable zones for tea production will shift, providing new benefits for some and huge risks for others. Facing these opportunities and threats in a climate-smart way will help farmers to proactively address climate change and ensure we all still get our daily infusion of liquid gold.

REFERENCES

1. Lehmann-Danzinger, H. Diseases and pests of tea: Overview and possibilities of integrated pest and disease management. *J. Agri. Trop. Subtrop.* **101**, 13–38 (2000).
2. Duncan, J., Saikia, S., Gupta, N. & Biggs, E. Observing climate impacts on tea yield in Assam, India. *Appl. Geogr.* **77**, 64–71 (2016).
3. Ritchie, H. Global tea production, 2014. *Ourworldindata.org.* https://ourworldindata.org/grapher/tea-production (2018).

4. Bhagat, R. *et al. Report of the Working Group on Climate Change of the FAO Intergovernmental Group on Tea.* Food and Agriculture Organization of the United Nations. http://www.fao.org/3/a-i5743e.pdf (2016).

5. Doublet, G. & Jungbluth, N. Life cycle assessment of drinking Darjeeling tea. *Conventional and Organic Darjeeling Tea* (ESU-services Ltd., Uster, 2010).

6. Elbehri, A. *et al. Kenya's Tea Sector Under Climate Change.* Food and Agriculture Organization of the United Nations. http://www.fao.org/3/a-i4824e.pdf (2015).

7. EST. *At Home with Water.* Energy Saving Trust, London, UK. http://www.energysavingtrust.org.uk/sites/default/files/reports/AtHomewith Water%287%29.pdf (2013).

8. WRAP. Household food and drink waste in the United Kingdom 2012. *Waste and Resource Action Programme.* http://www.wrap.org.uk/sites/files/wrap/hhfdw-2012-main.pdf.pdf (2013).

9. Rowlatt, J. The dark history behind India and the UK's favourite drink. *BBC Online.* https://www.bbc.co.uk/news/world-asia-india-36781368 (2016).

10. Xia, E.-H. *et al.* The tea tree genome provides insights into tea flavor and independent evolution of caffeine biosynthesis. *Mol. Plant* **10**, 866–877 (2017).

11. Kahn, B. *Climate Change Poses a Brewing Problem for Tea.* Climate Central. https://www.climatecentral.org/news/climate-change-altering-tea-industry-19071 (2015).

12. Assam reeling under heatwave, death toll 6. *Times of India.* https://timesofindia.indiatimes.com/city/guwahati/Assam-reeling-under-heatwave-death-toll-6/articleshow/20564314.cms (2013).

13. Bolton, D. 'Lowest Rainfall in Living Memory' Assam drought affects all. *World Tea News.* https://worldteanews.com/tea-industry-news-and-features/lowest-rainfall-living-memory-assam-drought-affects-tea (2014).

14. Darade, P. Monsoon 2017: Death toll rises in Gujarat, Assam floods. *India. com.* https://www.india.com/news/india/monsoon-2017-death-toll-rises-in-gujarat-assam-floods-landslide-in-himachal-pradesh-as-rains-continue-2331698/ (2017).

15. Singh, D., Tsiang, M., Rajaratnam, B. & Diffenbaugh, N. S. Observed changes in extreme wet and dry spells during the South Asian summer monsoon season. *Nat. Clim. Chang.* **4**, 456 (2014).

16. Kahn, B. Global warming changes the future for tea leaves. *Scientific American.* https://www.scientificamerican.com/article/global-warming-changes-the-future-for-tea-leaves/ (2015).

17. Roy, S., Muraleedharan, N., Mukhapadhyay, A. & Handique, G. The tea mosquito bug, Helopeltis theivora Waterhouse (Heteroptera: Miridae): Its status, biology, ecology and management in tea plantations. *Int. J. Pest Manage.* **61**, 179–197 (2015).

18. Ahmed, S. *et al*. Effects of extreme climate events on tea (Camellia sinensis) functional quality validate indigenous farmer knowledge and sensory preferences in tropical China. *PLoS One* **9**, e109126 (2014).

19. FAO. *Socio-Economic Implications of Climate Change for Tea Producing Countries*. Food and Agriculture Organization of the United Nations Intergovernmental Group on Tea. http://www.fao.org/fileadmin/templates/est/meetings/IGGtea21/14-4-ClimateChange.pdf (2014).

20. Carbonbrief. Mapped: How every part of the world has warmed—And could continue to warm. *Carbonbrief.org*. https://www.carbonbrief.org/mapped-how-every-part-of-the-world-has-warmed-and-could-continue-to-warm (2018).

21. Dutta, R. Climate change and its impact on tea in Northeast India. *J. Water Clim. Change* **5**, 625–632 (2014).

22. Borgohain, P. Occupational health hazards of tea garden workers of Hajua and Marangi tea estates of Assam, India. *Clarion* **2**, 129–140 (2013).

23. Kalita, R. M., Das, A. K. & Nath, A. J. Carbon stock and sequestration potential in biomass of tea agroforestry system in Barak Valley, Assam, North East India. *Int. J. Ecol. Environ. Sci.* **42**, 107–114 (2017).

24. Varshney, J. G. & Sushilkumar, M. J. *Proceedings of Taal2007: The 12th World Lake Conference* 1039–1045.

25. Tripathi, S. D. & Sharma, B. K. Integrated fish-duck farming. *Integrated Agriculture-Aquaculture. FAO Fisheries Technical Paper 407*. http://www.fao.org/docrep/005/Y1187E/y1187e14.htm (2001).

26. Chen, L., Apostolides, Z. & Chen, Z.-M. *Global tea breeding: Achievements, challenges and perspectives*. (Springer Science & Business Media, 2013).

27. Lu, C. Climate change threatens tea—But its DNA could save it. *CNN*. https://edition.cnn.com/2017/05/11/world/conversation-climate-tea-crisis/index.html (2017).

CHAPTER 5

Climate-Smart Milk

Abstract Global average milk consumption per person now tops 100 kilograms a year. In the US and across much of Europe we put away well over double this average. Some 700 million tonnes is produced worldwide annually with the UK being a net exporter of milk products and boasting a dairy herd numbering almost two million cows. Cows are a major source of the powerful greenhouse gas methane. Each litre of fresh milk we purchase is responsible for the equivalent of 3 kilograms of greenhouse emissions—over half a kilogram per standard glass. Milk is one of the most wasted foods in the UK at 290,000 tonnes each year. Reduced wastage by households is therefore central to reducing milks carbon footprint. On farms, improved animal health can provide major emissions benefits, as can alterations in feed. Higher temperatures pose a risk to yields and may increase milk spoilage. Fodder quality is also likely to be reduced in a future climate while some major diseases like Blue Tongue could benefit from warming and changing rainfall patterns.

Keywords Methane • Spoilage • Fodder quality • Methanogens • Blue Tongue • Liver flukes • Agroforestry • Heat waves • Ruminants • Antibiotic resistance

© The Author(s) 2019
D. Reay, *Climate-Smart Food*,
https://doi.org/10.1007/978-3-030-18206-9_5

Milk is a must-have for most breakfasts. Whether drunk neat, added to tea and coffee, or consumed as butter, cream, yoghurt and cheese, the global average consumption per person now tops 100 kilograms of the stuff each year. In the US and across much of Europe we put away well over double this average, though the global king of the milk moustache is Finland at over 400 kilograms of milk per person per year (Fig 5.1).

The vast majority of what we consume is in some form of cow's milk, with around 700 million tonnes produced worldwide annually. It can be produced in a very wide range of climates as long as the cattle are kept well fed. As a result, most regions of the world are deemed to be self-sufficient [2], and the fresh milk consumed in a country is often produced there too.

The UK is no exception to this, being a net exporter of milk products and boasting a dairy herd numbering almost two million cows [3]. Milk drinking here was institutionalised in schools and nurseries across post-war Britain, with the provision of free milk to all children up to the age of five (designed to give them a boost in nutrition at a time when food supplies

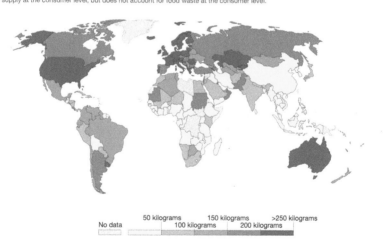

Per capita milk consumption, 2013

Average per capita milk consumption, measured in kilograms per person per year. This includes the milk equivalents of dairy products made from milk ingredients, but excludes butter. Data is based on per capita food supply at the consumer level, but does not account for food waste at the consumer level.

Our World in Data

No data   50 kilograms   150 kilograms   >250 kilograms
           100 kilograms   200 kilograms

Fig. 5.1 Global milk consumption per capita in 2013 (Source: Hannah Ritchie, Our World in Data) [1]. Available at: https://ourworldindata.org/grapher/per-capita-milk-consumption

were low). This school milk provision continued for decades and was extended to older school children too. For many it is still a fond memory and one that set them on the road to a lifetime of milk drinking. Personally, I hated it.

In the Britain of the 1970s cut backs had already meant that free milk supplies were reduced. Unfortunately for me, our small primary school in rural Lincolnshire still got its daily supply. Each morning a crate of small bottles with thin foil lids was delivered to outside the Headmaster's entrance. On cold days it froze—the milk pushing up the lids to leave short white lollies. On warm days it baked and was pecked at by wild birds. Whether frozen, pecked or near-rancid, each child was given a bottle and made to drink it. I was a hungry boy and constantly vying with my brothers for food at home. Yet a few fraught episodes of being forced to drink the warm, borderline-rancid milk of school put me off it for life.

As an estimated nine out of ten people in Britain still regularly consume fresh milk [4], most would seem to have avoided such hang-ups. British consumers bought some five and a half billion litres of fresh milk in 2017, at a cost of over £3 billion [5]. Milk remains a big global foodstuff. Unfortunately for our climate, it can also have a very big carbon footprint.

Cows are a major source of the powerful greenhouse gas methane. Most of this methane is formed in a cow's rumen (its first stomach) by microbes involved in the breakdown and fermentation of grass and other feeds. The bulk of it is then belched back into the atmosphere, with a single Daisy, the dairy cow, able to produce hundreds of litres of methane in a single day [6] (over 100 kilograms over the course of a year [7]). More emissions then arise from cattle manure and urine, from land use change and cattle feed production, and from the collection, processing and distribution of milk itself [8].

The result is that each litre of fresh milk we purchase is responsible for the equivalent of 3 kilograms of greenhouse emissions [9]—that's over half a kilogram per standard glass of the white stuff [8, 10]. For most consumers, their direct role in this hefty footprint might seem minor. Yes, there's the transport of milk from store to home, and the electricity we use to power our refrigerators, but these amount to less than a tenth of the total [8]. Where we can really dent the lifecycle emissions of milk (and in this case it's a very deep dent) is through less waste.

In the US, consumers waste around one-fifth of the milk they buy, with about 12 per cent also being wasted by retailers [8]. Likewise, milk is one of our biggest throw-away foods in the UK—290,000 tonnes each year at a cost to households of £290 million [11]. The reasons for such huge amounts of avoidable milk wastage centre on it not being used in time, too much being served and personal preference (fussy eaters like me refusing it). Consumers giving the milk they buy the best possible chance of being used is at the heart of tackling this waste [12]. This includes smaller, more frequent milk shopping, keeping it cold and knowing what you need—checking how much is in the fridge and its use-by date for instance [13].

That ripple effect of consumer action on food waste, then reducing emissions all the way back up through the supply chain, could be enormous for milk. Halving avoidable milk waste by UK households would deliver a cut equivalent to over 400,000 tonnes of emissions each year. For the US, comprehensively tackling milk waste by stores and households (cutting waste to just 3 and 5 per cent respectively) would together slice a quarter off the life cycle emissions of milk there [8] and deliver a staggering 8 million tonne reduction in the greenhouse gas emissions of the US dairy industry.

*   *   *

As climate change intensifies the importance of a resilient supply chain for high emission-short lifetime foods becomes even more important. There's a potential double-whammy in that the shelf lives of highly perishable foods like milk are themselves at risk—rising temperatures will increase food spoilage unless refrigeration is able to keep up. These cold-chain climate impacts have so far mainly been looked at in terms of food poisoning, such as accelerated growth of *Salmonella* on meat in warmer conditions [14]. For milk, the extra load put on refrigeration to keep it fresh in a future climate may well increase electricity demand, but any extra emissions from this would be dwarfed by the savings made through reduced spoilage and waste.

For many households in the developed world this may simply involve upgrading to a more energy-efficient fridge and making sure the food is not left out on worktops for too long. Further up the supply chain, more intense heat waves and the potential for transport delays due to floods will require everyone from farmers and processors to distributors and retailers to look at whether extra refrigeration, storage capacity and back-up power generation are needed [14].

Some manufacturers are trying to extend the lifetime of milk products using new technology, such as addition of bacterial strains that slow fungal growth and spoilage [15]. Longer-life milk is nothing new [16], with the use of ultra-high temperature (UHT) treatment having been around since the 1960s. This heat sterilisation means that, unopened, it can last without refrigeration for over six months. Condensed and sweetened milk has similar life-extending properties, usually coming in cans and often used instead of cream on deserts. As such, more uptake of these long-life milks could give more climate resilience to supplies and slash spoilage rates. A major downside to these types of milk is their distinctive 'cooked' taste.

My own uncle Bern loved the stuff, with no trip to the beach complete without a can of condensed Carnation milk to pour into his super-strong tea. Yet for those of us already unconvinced by the joys of fresh milk, this sweet and slightly brown-coloured stuff was horrible. As with so many foods, long-life milk is an acquired taste. It remains very popular in France, Spain and Italy where higher temperatures make it especially useful, while in the colder climes of Britain and Scandinavia fresh milk is preferred. For Uncle Bern, a pre-war childhood without fridges and an army career in the heat of North Africa guaranteed his life-long love for the dubious pleasures of condensed milk.

For most dairy farmers, preventing milk spoilage relies on strategies like good cattle husbandry, clean equipment and well-maintained chilling and storage systems—failing refrigeration can be a big source of milk spoilage insurance claims [17]. In less developed nations, limited access to refrigeration technology and reliable power supplies increases the risk of milk spoilage where temperatures soar or transport links are cut [18].

The sometimes-treacherous journey of milk from cow to our breakfast table is clearly one that could be made still more fraught by climate change. To date the supply chain risks of extreme weather have had relatively little attention worldwide. Instead, it is the very beginning of our milk's journey that has grabbed the most headlines and attracted the most research: the cows themselves.

Cows, like humans, don't like it too hot. If the temperature begins to push up into the high 20s Celsius (80s Fahrenheit) then heat stress impacts may start to show. First the cow becomes lethargic and sweaty, her breaths becoming shallower and faster. As temperatures move into the 30s Celsius she may start to pant and her production of milk plummet. Without relief from the heat she may die.

Modern dairy cows tend to be more susceptible to heat stress, as their high feed intakes, sizes and growth rates mean they generate more body heat [19]. Heat stress can also lead to a weakened immune system and the spread of diseases like mastitis [20]. Historically these kinds of hot weather impacts were only a big problem for cows in warm and tropical climates, but increasing temperatures at higher latitudes—including the US, Canada and Europe—have increased the risks to cattle there too [21].

Even where heat stress is relatively mild it can mean big reductions in milk production, and so major losses for farmers [22]. In the US alone it is estimated to cost the dairy industry $900 million a year through reduced milk and calves, and increased culling. Production losses and increased mortality rates caused by high temperatures have been documented in many US states over the last decade [23]. In the summer of 2017, it was California—the biggest milk-producing state in the country—that showed just how damaging extreme heat can be.

In late June that year, temperatures began to rise quickly and hit the mid-30s Celsius (110 Fahrenheit). Milk production in the worst affected regions slumped by a fifth and an estimated 4,000 cattle died [24]. Without adaptation, climate change will make such impacts more frequent, more widespread and more severe. It is predicted that excessive heat and humidity could cost the US over $2 billion a year by the end of the century if no action is taken. Hot, southern states like Florida may lose as much as a quarter of their milk production [25].

Over in Europe a similar trend of northward-creeping heat stress risk is emerging. The devastating heat wave of 2003 hit both humans and livestock hard, with 4,000 extra cattle deaths per week reported [26]. Subsequent heat waves in 2006 and in 2017 (the aptly-named 'Lucifer' heat wave) again took their toll on Europe's livestock and milk production [27].

For much of the twentieth century heat stress was a non-issue for most UK dairy herds; between 1973 and 2012 the average farm saw just 1 day each year where temperature and humidity climbed to a level where mild heat stress might occur [28]. The heat waves of the new millennium changed all that. Across the Midlands and the south of England, dairy cows typically experienced five days of heat stress-inducing conditions in the hot summers of 2003 and 2006. Milk yields from some herds were markedly reduced for a time, but these events served mostly as a warning of what the future may hold. Based on projected climate change in the UK

[29], dairy herds in the south could be enduring an average of 30 to 40 days of such heat stress conditions each year by the end of this century.

To date, a far more familiar severe weather risk for dairy farmers in the UK is that of heavy rainfall and flooding. Waterlogged soils may make pasture land less productive and much more vulnerable to damage by grazing. In very wet years farmers are forced to keep their herds indoors for longer, meaning more reliance on cattle feed. Availability and quality of forage and feed itself can also suffer in really wet years (like those of 2012 and 2013), with higher costs and lower milk yields as a result [30]. Indeed, in many areas of the world there is a risk that future climate change impacts combined with a carbon dioxide-enriched atmosphere will mean big reductions in the quality of forage [31].

Along with the wetter winters and more extreme rainfall events may come the less visible but far more dangerous threat of disease. Dairy cows are valuable animals and farmers will try all they can to keep them well. Despite precautions, diseases like Foot and Mouth have devastated herds in recent years [32]. In 2007, discovery of Blue Tongue in British livestock again caused widespread concern [33]. This viral disease is spread by midges and had emerged in North Western Europe the previous year. It is spread most rapidly in warm and wet conditions, with its 2006 outbreak being attributed to the warming that has occurred in this region over the past 50 years [34].

As climate changes, so the distribution of disease vectors like midges will also change. Increasing risks of Blue Tongue are now predicted across most of Europe, with a warming climate meaning its range may expand across the US, western Russia and central Africa too [35]. The strict quarantine, livestock tracking and monitoring regime now in place in the UK [36] has so far helped to lower the risk of outbreaks of such non-endemic diseases. For some already-established cattle diseases and parasites, however, climate change could make existing problems a whole lot worse.

Liver flukes are flat parasitic worms that mainly affect cattle and sheep. Even a light infection can damage liver function and reduce productivity—a heavy infection can kill the host animal. In the UK these parasites rely on a life cycle that starts off with eggs produced by the adult flukes in a cow's liver being excreted along with manure. If temperatures are high enough (over 10 degrees Celsius) the eggs develop quickly and produce the first microscopic mobile stage of the parasite. These then search out and infect the water snails common to many wet, low-lying grasslands. Within the snails the parasites grow and multiply fast (the warmer it is the faster they develop). After around 6 weeks the second mobile stage is released and these spread through the vegetation where they become infective cysts waiting for the next passing cow or sheep to chomp down on them [37].

As Britain has experienced warmer conditions and more flooding of grasslands, so the liver fluke parasites and their water snail vectors have flourished. More intense summer droughts have the potential to limit them in some areas in the future, but a trend of higher temperatures and more extreme rainfall risks enhancing the spread and impact of liver flukes across the UK [38, 39].

* * *

As a rule of thumb, where the welfare of dairy cows goes up greenhouse gas emissions come down. This certainly applies for adaptation to increasing heat—the majority of those livestock farmers in California who faced the wilting summer heat of 2017 have already put in place plans to protect their cows.

Many introduced shading in feeding, drinking and corral areas to give cows plenty of opportunities to seek respite from the sun when they need it [40]. Others use water sprayers and misters—as long as water supplies allow it—to cool the cows by evaporation. Some farms even employ large fans in the holding areas outside milking parlours, to keep air moving and temperatures down.

For those UK dairy farmers facing heat waves, similar sorts of cooling strategies could be the way forward. Here in Britain though, cattle are more often grazed on large areas of pasture during summer, and so may only come into more enclosed areas for milking times. At the milking parlour and its surrounds, providing shade, water sprays and fans can again reduce heat stress risks.

Proper refrigeration of the collected milk is also at risk with higher temperatures. Here, heat exchangers (that pre-cool the milk before refrigeration), and heat recovery units (that then collect the heat and use it for water heating), can radically reduce farm energy use and so greenhouse gas emissions [41].

Out in the fields there is often an opportunity to use the natural shading and shelter provided by trees to increase hot weather resilience—dairy cows given such shaded areas have shown reduced panting and heat stress symptoms [42]. Though, as we saw with Assam's tea gardens, the integration of trees with agriculture (agroforestry) is most commonly associated with growing crops, trees are a successful part of livestock systems around the world too [43]. Tree shelterbelts around fields can reduce the impacts of extreme weather events, including storms, intense rainfall, and extremes of heat and cold [44]. For areas of intensive agriculture they also represent an important opportunity to sequester more carbon dioxide from the atmosphere without having to lose productive farmland [45]. Some farmers have extended the benefits of livestock agroforestry to include extra forage for the animals, a source of biofuel for energy generation, and even as a natural filter for pollutants—the trees can help reduce nitrate leaching to drainage streams and capture ammonia emissions to the atmosphere [46, 47].

Just as heat wave warnings are now widely used to reduce risks to human health, so such warnings can help livestock farmers get plans in place to protect their herds. Reducing the numbers of cows held in confined spaces like milking parlours can be a good way to allow heat to disperse more easily. Likewise, providing additional drinking water supplies and shifting feeding times, so that cows are not all feeding during the hottest parts of the day, will cut heat stress risks. Even the cattle feed itself can be modified to make it more energy-dense and so reduce how much extra body heat is produced as it is digested [48]. In fact, changing what cows eat has a peculiar strand of food and climate change science all of its own.

Excess production of methane in the rumens of dairy cows is bad both for our climate and for dairy farmers. The microbes that produce the methane—methanogens—make use of the carbon dioxide and hydrogen generated as feed is fermented and digested. With harder-to-breakdown food, such as straw, more hydrogen is generated and so methane emissions tend to increase [49]. Providing dairy cows with higher quality feeds and forages can therefore mean less of the food is converted into methane and more of it into milk.

Scores of different feed and forage types have been assessed in terms of the methane penalties they incur [50]. Improving feed quality remains one of the standout strategies in efforts to boost production and reduce the carbon footprint of livestock. Yet, many farmers either do not have access to better feeds or their cattle range far and wide, making controlling what they eat near-impossible [51].

For those dairy farmers with closer control of the diet of their herds, and access to the latest feed mixes, there are some extra weapons in the methane-targeting armoury available [52]. While higher quality feeds shift digestion away from the hydrogen production the methanogens rely on, a host of feed additives can also be used to divert the hydrogen supply or even to target the methane-producing microbes themselves [53].

Australia is arguably the world leader in assessing exactly what might work best—their huge cattle and sheep populations mean livestock methane gets a lot of attention. The impacts of adding everything from tea, garlic and seaweed extracts [54], to cinnamon, curry spice and oregano have been examined over the years [55]. Many of these natural extracts work by directly inhibiting the methanogens. Others, like nitrate and sulphate additives, work by competing with the methanogens for any available hydrogen in the cow's rumen [56].

The results can be impressive—cuts in methane of over three-quarters have been reported in lab experiments [57]. They may also be short-lived. With prolonged exposure, the methane-producing microbes often become resistant to the effects of the additives. Too much use of nitrate additives can even prove toxic for the cows themselves [56]. The current front-runner as an additive that increases productivity, cuts emissions and boosts climate resilience—the holy climate-smart trinity—is not an exotic curry spice or a rare algal extract: it is fat [58].

Fats, especially those rich in fatty acids like sunflower oil, are able to reduce methane [59, 60] and the amount of heat generated during digestion [61]. These dietary fats can be derived from many natural sources, including algae. They also avoid many of the public health issues associated with artificial methanogen inhibitors like antibiotics—an antibiotic called monensin is widely used in livestock feed to boost growth and cut methane emissions, but is banned in Europe due to concerns around the spread of antibiotic resistance [62].

Where antibiotics have a less controversial role in delivering climate-smart milk is in fighting disease. Together with improved veterinary care and animal health extension services, access to livestock medicines can vastly increase resilience to diseases and parasites that would otherwise attack cattle [51, 63].

For dairy farmers in the UK there are now vaccines available to help control the midge-borne Blue Tongue virus, as well as regular warnings about new outbreak risks and tight livestock movement restrictions wherever infection is confirmed [64]. The march of parasites like the liver fluke has been slowed by the use of flukicide drugs that kill the parasites while they are inside the host animal [65]. Yet with more drug use has come more drug resistance. New vaccines may again hold the solution, giving protection in areas where drug resistance is already established and providing a longer-term defence for dairy farmers as liver flukes spread into new areas.

So, a healthy and happy cow is usually a more climate-resilient and lower-emissions cow. Even the happiest cows have a limit to the amount of milk they can produce though, and some breeds produce a lot more than others. For the carbon footprint of our breakfast milk, this is crucial.

In the push for bigger milk yields, dairy livestock have undergone intensive selection to emphasise sought-after traits. In just 60 years, genetic selection and improved management of North America dairy cows has quadrupled milk yields and halved the methane emissions of each litre of milk produced [60].

British herds are now dominated by the black and white Holstein-Friesian cow—a cross between the milk super-breed of Holstein imported from the US and Canada, and the fast-breeding Friesian cow that was the mainstay of UK dairy farming up until the 1980s. Holsteins are able to produce over 7 tonnes of milk a year (compared to around 6 tonnes per year for a Friesian) [66], so breeding cows that are all or mostly from Holstein stock makes sense for increasing milk yields. The downside is that more milk production may come at the cost of other desirable traits, like high fertility. The same large size and fast metabolism of Holstein cows that allows them to produce so much milk can also make them more susceptible to overheating and so more vulnerable to heat stress.

Further selection for and introduction of genetic traits [67]—like heat tolerance, higher yields or disease resistance—all have the potential to deliver climate-smart milk [60]. The real challenge is in finding the combination that works best for the specific locations and local circumstances of different dairy farms in a rapidly changing climate.

Beyond cow welfare and genetics, climate-smart milk relies on the whole dairy production system. If a new wonder feed wipes out livestock methane, but generates even bigger greenhouse gas emissions through its own production, then the climate benefit is lost. All cereals and crops have carbon footprints, so if they are then used to feed cattle this is added to the life cycle emissions of the milk we eventually drink [68]. In most cases though, cuts in dairy cow methane from improved feed will still outweigh the emissions from the feed itself. For many rangeland cows in the developing world the food they forage is wild-grown and inedible to humans. These browsing herds are effectively creating milk from 'zero carbon' feed, but often with hefty methane emissions in between thorny bush and milk churn, and so a big overall carbon footprint.

Cow manure is itself a globally important source of methane, while both it and cow urine are rich in nitrogen and so contribute to emissions of the powerful greenhouse gas nitrous oxide too [52]. For dairy farmers, improved manure and urine management can turn this animal waste burden into a climate-smart blessing. In areas where cows congregate (e.g. in cattle sheds and outside milking parlours) the waste can be collected. This avoids the risk of it being washed into drainage waters by heavy rain or emitting large amounts of ammonia to the air on hot days. The collected waste is then a valuable feedstock for anaerobic digestion—the deliberate production and capture of methane for use as an energy source. Many farms already do this, using the biogas to heat buildings, generate electricity, or even to pump into the wider gas supply network [69]. The residues from the anaerobic digester then make an excellent soil improver to apply back on the fields and substitute for artificial fertilisers.

Even where anaerobic digestion is not an option, separating the manure and urine into covered storage often reduces air and water pollution problems. Methane will still be produced though, and aerating the manure, reducing storage times or even destroying the methane by flaring, have all been suggested as ways of reducing its climate impact [52]. Cow diet can affect these waste emissions too. Ironically, the same nitrate supplements that inhibit gut methanogens may boost nitrous oxide production in the cow's manure and urine—potentially just swapping the climate change penalty of milk from one place and gas to another.

The final big opportunities for climate-smart milk on the farm come in the way manures and fertilisers are applied, and the ways the cows use their fields. Getting the timing and amounts of manure and fertiliser right maximises how much of the nitrogen it contains is used by the grass or crops, and so minimises losses to air and water. For cow behaviour, keeping them away from waterlogged areas and streams, regularly moving feeders and drinkers about, and placing field gates at the top of slopes (where it's usually drier) can all help to reduce the compaction and 'poaching' of soils, and so the pollution and greenhouse gas emissions that result.

Globally, the challenge of realising such climate-smart milk remains a daunting one. Through heat stress and disease, reduced food supply and quality, climate change is already damaging production and pushing up the emissions of the milk we consume. In the UK we are blessed with expert dairy farmers with access to many of the feed mixes, medicines, technologies and animal breeding programmes needed to deliver high milk yields that are both climate resilient and low carbon. Large parts of the developing world have none of this [68].

New programmes, such as those supported by the World Bank are starting to change things. In Burkina Faso, many farmers are already employing small biodigesters that produce methane from manure for home cooking. The switch from charcoal and wood fuel to biogas reduces deforestation and improves air quality. Applying the nutrient-rich residues to fields then helps to boost yields and improve water retention of the soils—vital as rainfall becomes less predictable. With financial help from 'carbon credits' gained by reducing greenhouse gas emissions, Burkina Faso's President is now aiming for 40,000 households to be using these biodigesters by 2020 [70].

Elsewhere, provision of research, extension services and opportunities to share good practice is helping to identify precisely which climate-smart milk strategies could be most feasible for specific situations. Large, intensive, Western-style dairy farming might look like a good goal on paper, but in reality this model risks distorting labour markets and damaging community cohesion if applied universally [51]. As ever, context-specific climate-smart solutions trump any amount of misplaced shiny technology.

REFERENCES

1. Ritchie, H. Global milk consumption per capita, 2013. *Ourworldindata.org.* https://ourworldindata.org/grapher/per-capita-milk-consumption (2018).
2. IDF. The World Dairy Situation 2016. *Bulletin of the International Dairy Federation 485.* https://www.idfa.org/docs/default-source/d-news/world-dairy-situationsample.pdf (2016).
3. AHDB. UK cow numbers. *AHDB Dairy.* https://dairy.ahdb.org.uk/market-information/farming-data/cow-numbers/uk-cow-numbers/#.XOqSEy2ZPUo (2018).
4. DairyUK. *The UK Dairy Industry.* https://www.dairyuk.org/the-uk-dairy-industry/ (2018).
5. AHDB. *Milk and Cream Market.* https://dairy.ahdb.org.uk/market-information/dairy-sales-consumption/liquid-milk-market/#.XOqSSy2ZPUo (2018).
6. Adam, D. How much brown cow? *Nature.* https://www.nature.com/news/1998/000907/full/news000907-6.html (2000).
7. Dong, H., Mangino, J., McAllister, T. & Have, D. Emissions from livestock and manure management. *2006 IPCC Guidelines for National Greenhouse Gas Inventories* (2006).

8. Thoma, G. *et al.* Greenhouse gas emissions from milk production and consumption in the United States: A cradle-to-grave life cycle assessment circa 2008. *Int. Dairy J.* **31**, S3–S14 (2013).

9. Poore, J. & Nemecek, T. Reducing food's environmental impacts through producers and consumers. *Science* **360**, 987–992 (2018).

10. Tesco. Product Carbon Footprint summary. https://www.tescoplc.com/assets/files/cms/Tesco_Product_Carbon_Footprints_Summary(1).pdf (2012).

11. WRAP. Household food and drink waste in the United Kingdom 2012. *Waste and Resource Action Programme.* http://www.wrap.org.uk/sites/files/wrap/hhfdw-2012-main.pdf.pdf (2013).

12. WRAP. Preventing household dairy waste. *WRAP Information Sheet—Courtauld Commitment 2.* http://www.wrap.org.uk/sites/files/wrap/Information sheet dairy _approved 12 11 10_.pdf (2010).

13. WRAP. The milk model: Simulating food waste in the home. *WRAP Final Report.* http://www.wrap.org.uk/sites/files/wrap/Milk Model report.pdf (2013).

14. James, S. & James, C. The food cold-chain and climate change. *Food Res. Int.* **43**, 1944–1956 (2010).

15. Westergaard-Kabelmann, T. & Olsen, M. D. Reducing food waste and losses in the fresh dairy supply chain. *Chr. Hansen Impact Study.* QBis Consulting. https://www.chr-hansen.com/_/media/203cf0800674489ab66c5acb666e923d.pdf (2016).

16. Rysstad, G. & Kolstad, J. Extended shelf life milk—Advances in technology. *Int. J. Dairy Technol.* **59**, 85–96 (2006).

17. FMG. *Milk Contamination and Spoilage.* FMG Risk Advice Guide. https://www.fmg.co.nz/globalassets/advice/fmgs-milk-contamination-risk-advice-guide_single.pdf (2015).

18. Liddiard, R., Gowreesunker, B. L., Spataru, C., Tomei, J. & Huebner, G. The vulnerability of refrigerated food to unstable power supplies. *Energy Procedia* **123**, 196–203 (2017).

19. Fidler, A. P. & VanDevender, K. *Heat Stress in Dairy Cattle.* University of Arkansas Extension FSA3040. https://www.uaex.edu/publications/pdf/fsa-3040.pdf.

20. AHDB. *Heat Stress in Cattle.* https://dairy.ahdb.org.uk/technical-information/animal-health-welfare/mastitis/working-arena-prevention-of-infection/housing/heat-stress-in-dairycattle/#.XOqSdC2ZPUo (2018).

21. Polsky, L. & von Keyserlingk, M. A. Invited review: Effects of heat stress on dairy cattle welfare. *J. Dairy Sci.* **100**, 8645–8657 (2017).

22. St-Pierre, N., Cobanov, B. & Schnitkey, G. Economic losses from heat stress by US livestock industries1. *J. Dairy Sci.* **86**, E52–E77 (2003).

23. Bishop-Williams, K. E., Berke, O., Pearl, D. L., Hand, K. & Kelton, D. F. Heat stress related dairy cow mortality during heat waves and control periods in rural Southern Ontario from 2010–2012. *BMC Vet. Res.* **11**, 291 (2015).

24. CBS. Cow carcasses pile up in California as heat wave causes mass death. *CBS News.* https://www.cbsnews.com/news/cow-carcasses-pile-up-in-california-as-heat-wave-causes-mass-death/ (2017).

25. Mauger, G., Bauman, Y., Nennich, T. & Salathé, E. Impacts of climate change on milk production in the United States. *Prof. Geogr.* **67**, 121–131 (2015).

26. Morignat, E. *et al.* Assessment of the impact of the 2003 and 2006 heat waves on cattle mortality in France. *PLoS One* **9**, e93176 (2014).

27. Henley, J. Extreme heat warnings issued in Europe as temperatures pass 40C. *The Guardian.* https://www.theguardian.com/world/2017/aug/04/extreme-heat-warnings-issued-europe-temperatures-pass-40c (2017).

28. Dunn, R. J., Mead, N. E., Willett, K. M. & Parker, D. E. Analysis of heat stress in UK dairy cattle and impact on milk yields. *Environ. Res. Lett.* **9**, 064006 (2014).

29. Carbonbrief. Mapped: How every part of the world has warmed—And could continue to warm. *Carbonbrief.org.* https://www.carbonbrief.org/mapped-how-every-part-of-the-world-has-warmed-and-could-continue-to-warm (2018).

30. Morris, J. Weather spells hard times for UK dairy farmers. *BBC Online.* https://www.bbc.co.uk/news/uk-england-devon-20532875 (2012).

31. Dumont, B. *et al.* A meta-analysis of climate change effects on forage quality in grasslands: Specificities of mountain and Mediterranean areas. *Grass Forage Sci.* **70**, 239–254 (2015).

32. Bates, C. When foot-and-mouth disease stopped the UK in its tracks. *BBC Online.* https://www.bbc.co.uk/news/magazine-35581830 (2016).

33. BBC. Bluetongue declared an outbreak. *BBC Online.* http://news.bbc.co.uk/1/hi/uk/7018205.stm (2007).

34. Guis, H. *et al.* Modelling the effects of past and future climate on the risk of bluetongue emergence in Europe. *J. R. Soc. Interface* **9**, 339–350 (2011).

35. Samy, A. M. & Peterson, A. T. Climate change influences on the global potential distribution of bluetongue virus. *PLoS One* **11**, e0150489 (2016).

36. DEFRA. *Beef Cattle and Dairy Cows: Health Regulations.* Department for Environment, Food & Rural Affairs, UK. https://www.gov.uk/guidance/cattle-health (2012).

37. AHDB. *Sustainable Worm Control Strategies for Cattle.* AHDB Dairy. https://dairy.ahdb.org.uk/non_umbraco/download.aspx?media=16159 (2013).

38. Gale, P., Drew, T., Phipps, L., David, G. & Wooldridge, M. The effect of climate change on the occurrence and prevalence of livestock diseases in Great Britain: A review. *J. Appl. Microbiol.* **106**, 1409–1423 (2009).

39. Fox, N. J. *et al.* Predicting impacts of climate change on Fasciola hepatica risk. *PLoS One* **6**, e16126 (2011).
40. Tresoldi, G., Schütz, K. & Tucker, C. Cow cooling on commercial drylot dairies: A description of 10 farms in California. *California Agri.* **71**, 249–255 (2017).
41. FCCT. Energy efficiency advice for dairy farmers. *Farmcarbontoolkit.org.uk.* https://farmcarbontoolkit.org.uk/toolkit/energy-efficiency-advice-dairy-farmers (2018).
42. Schütz, K., Cox, N. & Tucker, C. A field study of the behavioral and physiological effects of varying amounts of shade for lactating cows at pasture. *J. Dairy Sci.* **97**, 3599–3605 (2014).
43. Dawson, I. K. *et al. Agroforestry, Livestock, Fodder Production and Climate Change Adaptation and Mitigation in East Africa: Issues and Options* (World Agroforestry Center, Nairobi, Kenya, 2014).
44. Smith, J., Zaralis, K., Gerrard, C. & Padel, S. Agroforestry for livestock systems. *SOLID Farmer Handbook: Technical Note 12.* https://www.agricology.co.uk/resources/agroforestry-livestock-systems (2016).
45. Feliciano, D., Ledo, A., Hillier, J. & Nayak, D. R. Which agroforestry options give the greatest soil and above ground carbon benefits in different world regions? *Agri. Ecosyst. Environ.* **254**, 117–129 (2018).
46. Bealey, W. *et al.* The potential for tree planting strategies to reduce local and regional ecosystem impacts of agricultural ammonia emissions. *J. Environ. Manag.* **165**, 106–116 (2016).
47. Wilson, M. & Lovell, S. Agroforestry—The next step in sustainable and resilient agriculture. *Sustainability* **8**, 574 (2016).
48. Thomas, C. *Seven Practical Heat Stress Abatement Strategies.* Michigan State University Extension. https://www.canr.msu.edu/news/seven_practical_heat_stress_abatement_strategies (2012).
49. Yáñez-Ruiz, D. R. *et al. Feeding Strategies to Reduce Methane and Ammonia Emissions.* EIP-AGRI Focus Group Reducing Livestock Emissions from Cattle Farming. https://ec.europa.eu/eip/agriculture/sites/agri-eip/files/fg18_mp_feeding_strategies_2017_en.pdf (2017).
50. Iqbal, M. F., Cheng, Y.-F., Zhu, W.-Y. & Zeshan, B. Mitigation of ruminant methane production: Current strategies, constraints and future options. *World J. Microbiol. Biotechnol.* **24**, 2747–2755 (2008).
51. Ericksen, P. J. & Crane, T. A. The feasibility of low emissions development interventions for the East African livestock sector: Lessons from Kenya and Ethiopia. *ILRI Research Report* 46 (2018).
52. Gerber, P. *et al.* Technical options for the mitigation of direct methane and nitrous oxide emissions from livestock: A review. *Animal* 7, 220–234 (2013).
53. Frater, P. *Feed Additives in Ruminant Nutrition.* Agriculture and Horticulture Development Board (AHDB). https://beefandlamb.ahdb.org.uk/wp-con-

tent/uploads/2013/04/Feed-additives-in-ruminant-nutrition-FINAL.pdf (2014).

54. Meale, S. *et al.* Including essential oils in lactating dairy cow diets: Effects on methane emissions1. *Anim. Prod. Sci.* **54**, 1215–1218 (2014).

55. Patra, A. K., Kamra, D. N. & Agarwal, N. Effects of extracts of spices on rumen methanogenesis, enzyme activities and fermentation of feeds in vitro. *J. Sci. Food Agri.* **90**, 511–520 (2010).

56. Yang, C., Rooke, J. A., Cabeza, I. & Wallace, R. J. Nitrate and inhibition of ruminal methanogenesis: Microbial ecology, obstacles, and opportunities for lowering methane emissions from ruminant livestock. *Front. Microbiol.* **7**, 132 (2016).

57. Durmic, Z. *et al.* In vitro screening of selected feed additives, plant essential oils and plant extracts for rumen methane mitigation. *J. Sci. Food Agri.* **94**, 1191–1196 (2014).

58. Grainger, C. & Beauchemin, K. Can enteric methane emissions from ruminants be lowered without lowering their production? *Anim. Feed Sci. Technol.* **166**, 308–320 (2011).

59. Rasmussen, J. & Harrison, A. The benefits of supplementary fat in feed rations for ruminants with particular focus on reducing levels of methane production. *ISRN Vet. Sci.* **2011** (2011).

60. Knapp, J., Laur, G., Vadas, P., Weiss, W. & Tricarico, J. Invited review: Enteric methane in dairy cattle production: Quantifying the opportunities and impact of reducing emissions. *J. Dairy Sci.* **97**, 3231–3261 (2014).

61. Knapp, D. & Grummer, R. R. Response of lactating dairy cows to fat supplementation during heat stress. *J. Dairy Sci.* **74**, 2573–2579 (1991).

62. EC. Ban on antibiotics as growth promoters in animal feed enters into effect. *European Commission Regulation 1831/2003/EC.* http://europa.eu/rapid/press-release_IP-05-1687_en.htm (2005).

63. Assan, N. Possible impact and adaptation to climate change in livestock production in Southern Africa. *IOSR J. Environ. Sci. Toxicol. Food Technol.* **8**, 104–112 (2014).

64. DEFRA. *GB Bluetongue Virus Disease Control Strategy.* Department for Environment, Food & Rural Affairs, UK. https://assets.publishing.service.gov.uk/government/uploads/system/uploads/attachment_data/file/343402/bluetongue-control-strategy-140727.pdf (2014).

65. Stevenson, C., Mahoney, R., Fisara, P., Strehlau, G. & Reichel, M. The efficacy of formulations of triclabendazole and ivermectin in combination against liver fluke (Fasciola hepatica) and gastro-intestinal nematodes in cattle and sheep and sucking lice species in cattle. *Aust. Vet. J.* **80**, 698–701 (2002).

66. HolsteinUK. Cow facts. *UKCows.com.* http://ukcows.com/holsteinUK/publicweb/Education/HUK_Edu_DairyCows.aspx?cmh=66 (2018).

67. Hansen, P. *49th Florida DairyProduction Conference*, Gainesville, 7pp.

68. Gerber, P. J. *et al. Tackling Climate Change Through Livestock: A Global Assessment of Emissions and Mitigation Opportunities* (Food and Agriculture Organization of the United Nations (FAO), 2013).

69. Harvey, D. Powered by poo: Somerset dairy farm enjoys biogas boom. *BBC Online.* https://www.bbc.co.uk/news/uk-england-somerset-35482839 (2016).

70. World Bank. Carbon credits serve up clean cooking options for West African Farmers. *WorldBank.org.* http://www.worldbank.org/en/news/feature/2018/03/06/carbon-credits-serve-up-clean-cooking-options-for-west-african-farmers?CID=CCG_TT_climatechange_EN_EXT (2018).

PART II

Break Time

CHAPTER 6

Climate-Smart Chocolate

Abstract Ghana, Ivory Coast and Indonesia are the major cocoa producers, with smallholder farmers in West Africa being responsible for around 60 per cent of global supply. Each year in the UK alone we consume around 500,000 tonnes of chocolate as bars and drinks, in cakes and biscuits. The average 40-gram bar of milk chocolate will carry with it a carbon footprint of around 200 grams. In the UK an estimated 18,000 tonnes of chocolate and sweets are wasted each year, responsible for around 90,000 tonnes of greenhouse gas emissions. High temperatures and drought can have severe impacts on cocoa yields, with Ghana likely to see suitable areas for cocoa pushed south as more northern areas become drier and hotter. Diseases like cocoa swollen shoot virus are already a major concern with around 300 million trees thought to be infected. Rehabilitation and renovation of existing cocoa plantations is at the heart of building resilience to climate change and securing the millions of livelihoods that depend on cocoa.

Keywords Ghana • Cote D'Ivoire • Cocoa • Agroforestry • Renovation • Rehabilitation • Disease resistance

Chocolate. A joy. To unwrap a smooth bar at break time and feel that boost in mood and energy [1]. As a child I was obsessed with the stuff. It was a rare treat with Christmas and Easter being the delectable chocolate oases that shimmered in my dreams for months in advance. On 1 December each year an Advent calendar would be ceremonially pinned to the kitchen wall, with a small chocolate hidden behind each cardboard door. Having three siblings meant the privilege of opening a door, and eating the glossy treat behind, fell every four days; six chocolates in all for Advent and every one of them savoured as a precious delicacy. Christmas Day itself meant still more chocolatey delights to be hoarded and bartered with. The importance of chocolate to these family festivals was most apparent when it was absent. One tearful Easter Sunday our mum brought down the carefully hidden box containing that year's longed-for chocolate eggs. Mice had eaten everything but the foil wrappers!

Chocolate and its key ingredient—cocoa—remains a beloved foodstuff around the world. As bars and biscuits, in cakes and drinks, we collectively consume over 7 million tonnes of chocolate each year in a global market worth around $100 billion [2]. Western nations lead the way in terms of amounts consumed, with the US at 5.5 kilograms per person per year, and the British, Germans and Swiss vying for global chocoholic title at over 8 kilograms per person [3]. None of these major chocolate-consuming nations have a climate suitable for cocoa production.

Cocoa is produced by the seeds and pods of a tree called *Theobroma cacao*. Native to the tropical zone of Central and South America, chocolate's journey to become today's global food essential can be traced back to 400 BC as a drink of Mayan traders in Costa Rica [4]. Apparently it took many years to catch on in the West as, without something to sweeten it, it tasted horrible.

Cocoa trees are now grown in a wide tropical belt spanning central and southwest Africa, South and East Asia, and much of South America. They require average temperatures of 18 to 30 degrees Celsius and, most importantly, plentiful rainfall throughout the year [4]. Ghana, Ivory Coast and Indonesia are the major producers, with smallholder farmers in West Africa being responsible for around 60 per cent of global supply (Fig. 6.1) [5].

The chocolate that enfolds our own Scottish break-time treat started life as cocoa beans in Ghana. These beans are used to produce both cocoa butter and cocoa liquor, with on-farm emissions (mainly due to the use of nitrogen fertilisers) and the processing phase (due to diesel use for roasting beans) being the main sources of greenhouse gas from the chocolate industry in Ghana [7].

Cocoa bean production, 2014
Annual cocoa bean production, measured in tonnes per year.

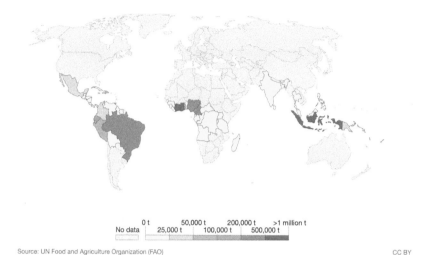

| | No data | 0 t | 50,000 t | 200,000 t | >1 million t |
| | | 25,000 t | 100,000 t | 500,000 t | |

Source: UN Food and Agriculture Organization (FAO)   CC BY

Fig. 6.1 Global cocoa bean production in 2014 by country of origin (Source: Hannah Ritchie, Our World in Data) [6]. Available at: https://ourworldindata. org/grapher/cocoa-bean-production

Chocolate's overall carbon footprint is an international pick 'n' mix of this initial cocoa production and processing, its transport around the world and the extra emissions due to the milk, sugar, palm oil and other ingredients that are then added to make the chocolate we know and love. The average 40-gram bar of milk chocolate will carry with it a carbon footprint of around 200 grams (or upwards of 5 kilograms per kilogram of chocolate). This rises to nearer 300 grams for a bar of dark chocolate due to the extra cocoa it contains.

Our chocolate, then, is a lovely but high-emission food. Each year in the UK alone we consume around 500,000 tonnes of it as bars and drinks, in cakes and biscuits [3]. Some 8 million of us eat it every day of the week and bars of solid milk chocolate are the nation's favourite [8]. The associated climate impact of British chocolate addiction is similarly huge, at around 2.5 million tonnes of greenhouse gas per year.

Chocolate is precious stuff, yet remarkably a significant amount of it never gets eaten. In the UK an estimated 18,000 tonnes of chocolate and sweets are wasted each year [9]. This avoidable waste is mainly due to it not being eaten before the use by date, with personal preference and spoilage accounting for the rest. Such avoidable chocolate wastage is responsible for around 90,000 tonnes of greenhouse gas emissions.

Avoiding waste is an obvious way for we consumers to reduce chocolate's climate impact, but our buying habits can also help to determine the sustainability of its global supply chain [10]. Demand has soared in recent decades, bringing increased rates of deforestation and land clearance in many cocoa-growing regions. This is adding to the already carbon–intensive nature of the chocolate life cycle. As emissions rise and climate change intensifies, chocolate is itself in the firing line when it comes to impacts. From cocoa farmers and processors right through to shops and consumers, making chocolate climate-smart may be the only way to secure the future of this most luxurious of foods.

Production of cocoa in its limited climate envelope of consistent rainfall and tropical temperatures has been pushed to its limits by surging demand and prices. The cash-crop that is chocolate has brought vital income and employment to many areas of West Africa and South America, yet increasing climate-variability threatens to undermine this production and so the welfare of the people who rely on it. In Cameroon in 2012, the Government announced plans to more than double cocoa production to 600,000 tonnes by 2020, providing a huge boost for farmer income and employment. The weather had other ideas.

Instead of reliable rains to recharge soils in the wet season and a dry season in which to harvest and dry the cocoa beans, the rains became more erratic [11]. Unexpected rainfall killed new flowers and blackened young pods, with the pesticides applied to the trees washed away. Farmers were forced to try and smoke-dry their beans and then face roads to market that had become rivers of mud. The unpredictable rains have also meant increased risks of pest and disease attacks—higher rainfall and humidity encouraging fungal diseases like Black Pod [12]. More than one-third of the cocoa harvest in the 2014–15 season was lost due to pests and diseases [11].

Over in Ghana, unreliable rainfall has caused big problems too. In the 2015–16 season the parching Harmattan wind that blows across Ghana from the Sahara Desert came early. With little rainfall and persistent moisture-sapping winds, the cocoa plants suffered poor growth and withering of pods in many areas [13]. Things could have been much worse. Given the huge importance of cocoa production in Ghana and the risks to it posed by severe weather events, its Government is already striving to boost resilience. Provision of early-maturing cocoa varieties and improved access to fertilisers have allowed farmers to increase production even in the face of more unreliable rainfall. Importantly, they and other big cocoa-growing nations, like Cote D'Ivoire, are working to understand the risks posed by climate change and help protect the longer-term future of the world's favourite treat.

* * *

The prime chocolate belt of cocoa production in West Africa stretches along the coast from Sierra Leone in the west through Cote D'Ivoire, Ghana and Nigeria, to Cameroon in the East. It employs around two million people and provides the world with almost three-quarters of the cocoa we crave. Most is still grown by smallholders who often have to contend with low cocoa prices, ageing trees (that give lower yields), poor access to technology and fertilisers, plus the weather vagaries of a changing climate. Over the last half-century, some areas have already experienced a drop in rainfall of almost one-third, making cocoa production there near impossible. With increasing temperatures and further changes in rainfall due to climate change in coming decades [14], more regions—including some farms in the key cocoa nations of Ghana and Cote D'Ivoire—are at risk of falling yields or even complete loss of cocoa production [15].

By the middle of this century annual rainfall across West Africa is not expected to change very much. Precipitation may even increase and the length of the dry season contract in some of the driest areas—giving a potential boost to cocoa trees. It is water availability, however, not rainfall, that is crucial for the plants. As temperatures rise, so water lost through evaporation increases. For some of the driest cocoa-growing areas in West Africa this means that, even with more rainfall, the amount of water actually available to the plants is reduced and drought impacts become more likely. By the middle of the century the climate envelope for cocoa is expected to contract southwards as drier, savannah-type conditions push down from the north. For big producers like Ghana, growers in the south should still do fine, but in northern areas across the West African chocolate belt a warmer future is set to make cocoa farming a whole lot harder.

Higher temperatures may also have a direct impact on cocoa. Cocoa plants grow best at between in the low to mid-20s Celsius. They can withstand temperatures of up to 38 degrees Celsius, but such high temperatures slow plant growth and damage pod development. As global temperatures increase, so does the risk that extreme heat wave events will cut cocoa yields in West Africa [15].

Predicting the impacts of climate change on chocolate production means understanding more than just the future climate envelope where growing cocoa will be possible. The farming systems and tree types, the global markets and competition, and the cocoa workers themselves, all are crucial in determining just how big a risk climate change poses. For the two million or so smallholder cocoa growers in Ghana and Cote D'Ivoire, average production currently reaches just half of its potential. Drops in productivity or market price can therefore put huge numbers of people at risk—over six million people in Ghana alone rely in some way on the cocoa industry. One of the reasons for low cocoa yields in Ghana is the old age of many of its cocoa trees. Yields peak at 18 years of age, and almost a quarter of the cocoa trees in Ghana are over 30 years old [16].

Widespread infection by diseases, such as the Cocoa Swollen Shoot Virus, is another concern. This disease is spread via mealy bugs that feed on the cocoa plants. It is thought that it was first transferred into cocoa plants when natural forests were cleared and its mealy bug vectors adopted the cocoa trees as their new hosts [17]. Swollen shoot virus causes leaves to become blotchy and lose their colour. As the infection spreads, the leaves sometimes develop red veins, the shoots and roots develop swellings, and the new shoots begin to die [18]. With severe infections cocoa yields can plummet by as much as 70 per cent and it can kill the whole tree within two or three years of infection [17]. Around 300 million cocoa trees are now thought to be infected, with the only effective treatment available to most farmers being complete destruction (burning) of the diseased trees and replanting with new, disease-free stock [19].

So, the future impacts of climate change on cocoa production—like higher temperatures and lower water availability in some areas of West Africa—are set to compound existing issues like old trees and endemic disease. This bitter recipe risks the livelihoods of millions of people and heaps extra pressure on land use—driving further deforestation and carbon emissions. Deforestation rates in Ghana and Cote D'Ivoire are some of the highest in the world, with over two million hectares of forest having been cleared for cocoa between 1988 and 2007 [5]. Unless cocoa farming in West Africa can be made more climate-resilient, our global chocolate addiction will mean even more climate-risk.

* * *

Climate-smart chocolate is already becoming a reality in Ghana and Cote D'Ivoire. Our own buying choices and greater awareness of the deforestation risks of cocoa production are helping to alter the practices of big chocolate companies and their suppliers [20]. More attention and support are being focused on the legion of smallholders that are the backbone of the cocoa industry. Here, helping farmers to increase productivity on their existing land, so as to avoid clearance of forest for new planting, is a prime focus. Firstly, rehabilitation of the existing cocoa trees through improving soils, fertiliser use, and better pest and disease control aims to boost their yields. So-called renovation of the cocoa is also encouraged, with replacement of old trees with younger disease-free ones that can gradually improve overall yields and give greater resilience in the future.

A big step in driving such change has been the growing commitment by the global chocolate industry to put an end to deforestation associated with the cocoa they use. Over 60 per cent of the world's cocoa is now supplied via companies with some level of commitment to tackle deforestation. To date, the success of this in actually curtailing deforestation has been questioned [21], yet the momentum for change is growing. Importantly, major chocolate players like Nestle, Hershey and Mars, and big retailers like Tesco and Marks & Spencer have signed up to the Cocoa & Forests Initiative [22]. Alongside national governments, this initiative sets out a framework within which all these cocoa stakeholders work both to end deforestation and to restore already degraded forest areas.

It is estimated that almost 2 million hectares of cocoa-growing land in Ghana and Cot D'Ivoire would benefit from rehabilitation—an area roughly the size of Belgium, with a further 1.3 million hectares likely needing some level of renovation [5]. Providing the huge number of good quality, disease-free cocoa seedlings that are now required across Ghana and Cote D'Ivoire is a big challenge. Many small holders don't have access to these (or don't know they do), even if they wanted to start replacing old trees for new ones right away. Where new seedlings have been supplied, there have been problems with mislabelling, damage in transport and delivery at times of the year when the new seedlings have little chance of success.

Ghana's government-owned cocoa board (Cocobod) controls all seed and seedling production, but its capacity is limited and it still has a long way to go before it can deliver replanting in the estimated one million hectares that need it. Ramping up such provision, by expanding cocoa plant nurseries and improving seedling distribution systems, has the potential to accelerate progress towards more productive and resilient cocoa growing.

For those areas at most risk from heat and drought impacts under a changing climate, the supply of new seedlings to farmers will need to extend beyond just healthy cocoa plants. Here, provision of other trees and plants that can be used to provide smallholders with shade for their cocoa, a source of timber, or to give an extra crop—like avocados—will become increasingly important [23]. The practice of growing shade trees alongside cocoa in Ghana has expanded in recent years as more farmers have experienced first-hand the threats posed by extreme drought and heat. Such shade cocoa can also reduce weed growth and, if combined with nitrogen-fixing trees, boost soil fertility. Its benefits are not a given though. By definition, shade cocoa means less light reaches the cocoa trees. In dry regions this can give vital protection from extreme heat and create a better microclimate in which the cocoa pods can develop. In wetter regions, where such heat and drought resilience are less important, the climate benefits of shade cocoa may be outweighed by slower growth rates, an increase in some diseases (like Black Pod disease), and reduced yields.

On some farms, even with rehabilitation, renovation and shade trees, cocoa's days are numbered. In these drier, already-marginal cocoa areas many smallholders now grow food crops, such as maize and vegetables, in rotation with their cocoa to supplement incomes. As the viable climate envelope for cocoa shifts away southwards, so such smallholders will need help to further diversify into non-cocoa livelihoods.

Knowing where and when such shifts will occur is vital for planning and pro-active adaptation. Really precise climate projections remain some way off—the climate models still struggle to deal with the kind of local scales and specific timings most relevant to smallholder cocoa farmers. Instead, information on larger scale, longer-term changes can be used to inform government plans and wider climate-smart strategies.

For the cocoa farmers themselves, addressing the current dearth of climate services, like provision of education, finance and training, is a prime target for helping to deliver widespread climate-smart improvements. Access to weather forecasts and climate information can reduce some of the risks posed by droughts and heat waves. Likewise, advice and support on choices of cocoa tree, how to best cultivate and prune them, and the ways in which to use pesticides (both to protect from disease and to avoid pollution), can improve productivity and resilience. Initiatives like Farmer Field Schools—where farmers share good practice and can access such advice and support [24]—have already proved successful in boosting the skills, yields and livelihoods of Ghana's farmers [25].

The huge scale of change required to deliver a productive, climate-smart cocoa future for Ghana requires a similarly large step-change in farmer knowledge and training. It also requires money. Most cocoa farmers in Ghana cannot access finance and, therefore, often cannot afford the upfront costs and gaps in income that arise from replanting or diversifying to other crops. Unlocking cheaper finance for these farmers, such as through government support, financial training and mobile phone technologies, could help to unblock this major bottleneck for sustainable cocoa across West Africa.

References

1. Nehlig, A. The neuroprotective effects of cocoa flavanol and its influence on cognitive performance. *Br. J. Clin. Pharmacol.* 75, 716–727 (2013).
2. Statista.com. *Retail Consumption of Chocolate Confectionery Worldwide from 2012/13 to 2018/19 (in 1,000 Metric Tons).* https://www.statista.com/statistics/238849/global-chocolate-consumption/ (2018).
3. Mintel. *Sweet Success for Seasonal Chocolate.* http://www.mintel.com/press-centre/food-and-drink/sweet-success-for-seasonal-chocolate (2017).
4. ICCO. *Origins of Cocoa and Its Spread Around the World.* International Cocoa Organization. https://www.icco.org/about-cocoa/growing-cocoa.html (2013).
5. Kroeger, A., Koenig, S., Thomson, A. & Streck, C. *Forest- and Climate-Smart Cocoa in Côte d'Ivoire and Ghana: Aligning Stakeholders to Support Smallholders in Deforestation-Free Cocoa* (World Bank, 2017).
6. Ritchie, H. Global cocoa bean production, 2014. *Ourworldindata.org.* https://ourworldindata.org/grapher/cocoa-bean-production (2018).
7. Ntiamoah, A. & Afrane, G. *Appropriate Technologies for Environmental Protection in the Developing World* 35–41 (Springer, 2009).
8. Mintel. *Nation of Chocoholics: Eight Million Brits Eat Chocolate Every Day.* http://www.mintel.com/press-centre/food-and-drink/nation-of-chocoholics-eight-million-brits-eat-chocolate-every-day (2014).
9. WRAP. Household food and drink waste in the United Kingdom 2012. *Waste and Resource Action Programme.* http://www.wrap.org.uk/sites/files/wrap/hhfdw-2012-main.pdf.pdf (2013).

10. kakaoplattform.ch. *Joint Forces for a Sustainable and Attractive Cocoa Sector.* https://www.kakaoplattform.ch/en/ (2018).
11. Ngalame, E. N. Extreme weather threatens Cameroon's hopes of becoming a cocoa giant. *Reuters.* https://www.reuters.com/article/us-cameroon-climatechange-cocoa/extreme-weather-threatens-cameroons-hopes-of-becoming-a-cocoa-giant-idUSKBN18Y1ON (2017).
12. Nkobe, M. K., Mulua, S. I., Armathée, A. J. & Ayonghe, S. N. Impacts of climate change and climate variability on cocoa (*Theobroma cacao*) yields in meme division, south west region of Cameroon. *J. Cameroon Acad. Sci.* **11** (2013).
13. Kpodo, K. Harsh winds, lack of rain to hit Ghana cocoa output. *Reuters.* https://www.reuters.com/article/ghana-cocoa-harmattan/harsh-winds-lack-of-rain-to-hit-ghana-cocoa-output-idUSL8N15N3QR (2016).
14. Carbonbrief. Mapped: How every part of the world has warmed—And could continue to warm. *Carbonbrief.org.* https://www.carbonbrief.org/mapped-how-every-part-of-the-world-has-warmed-and-could-continue-to-warm (2018).
15. Schroth, G., Läderach, P., Martinez-Valle, A. I., Bunn, C. & Jassogne, L. Vulnerability to climate change of cocoa in West Africa: Patterns, opportunities and limits to adaptation. *Sci. Total Environ.* **556**, 231–241 (2016).
16. Kongor, J. E. *et al.* Constraints for future cocoa production in Ghana. *Agroforest. Syst.*, 1–13 (2017).
17. Cilas, C., Goebel, F.-R., Babin, R. & Avelino, J. *Climate Change and Agriculture Worldwide* 73–82 (Springer, 2016).
18. Dzahini-Obiatey, H. & Fox, R. Early signs of infection in Cacao Swollen Shoot Virus (CSSV) inoculated cocoa seeds and the discovery of the cotyledons of the resultant plants as rich sources of CSSV. *Afr. J. Biotechnol.* **9** (2010).
19. Andres, C. *et al.* Agroforestry systems can mitigate the severity of Cocoa Swollen Shoot Virus disease. *Agric. Ecosyst. Environ.* **252**, 83–92 (2018).
20. Ionova, A. How fair is our food? Big companies take reins on sourcing schemes. *Reuters.* https://www.reuters.com/article/us-food-fairtrade-sustainability-insight/how-fair-is-our-food-big-companies-take-reins-on-sourcing-schemes-idUSKCN1BE0GI (2017).
21. Maclean, R. Chocolate industry drives rainforest disaster in Ivory Coast. *The Guardian.* https://www.theguardian.com/environment/2017/sep/13/chocolate-industry-drives-rainforest-disaster-in-ivory-coast (2017).
22. Nieburg, O. The final cut: What can the chocolate industry really do to halt cocoa deforestation? *Confectionery News.* https://www.confectionerynews.com/Article/2017/09/27/Cocoa-deforestation-What-can-the-chocolate-industry-really-do (2017).

23. Abdulai, I. *et al.* Characterization of cocoa production, income diversification and shade tree management along a climate gradient in Ghana. *PLoS One* **13**, e0195777 (2018).
24. FAO. *Global Farmer Field School Platform.* http://www.fao.org/farmer-field-schools/en/?utm_source=twitter&utm_medium=social+media&utm_campaign=faoknowledge (2018).
25. Okorley, E. L., Adjargo, G. & Bosompem, M. The potential of farmer field school in cocoa extension delivery: A Ghanaian case study. *J. Int. Agri. Exten. Edu.* **21**, 32–44 (2014).

CHAPTER 7

Climate-Smart Bananas

Abstract Some 130 countries grow bananas, with the 120 million tonnes they produce each year mainly being grown by smallholders for home consumption or sale at local markets. Around 800,000 tonnes are imported to the UK each year with each banana having a carbon footprint of 100–200 grams. We waste over a million bananas a day in Britain, at a cost to the climate of over 30,000 tonnes of emissions annually. Disease is the number one threat as almost all our bananas come from a single variety called Cavendish. A fungal disease called Topical Race 4 is already destroying large numbers of banana plants around the world. Storm damage, floods and drought likewise pose an increasing risk. Boosting plant health through good water and nutrient management can help to give resilience to climate change and disease threats alike. Organic production and the use of biological pest control are also proving successful.

Keywords Agroforestry • Banana-coffee • Hurricanes • Dominican Republic • Organic food • Tropical Race 4 • Cavendish banana • Fusarium • Panama disease • Black Sigatoka

Once so rare in the West they were considered an exotic fancy, the banana is now a staple part of the daily menu for millions. Whether you like yours greenish and firm, spotty and sweet, or mostly black and super-soft, the different ripening stages of the bananas in our fruit bowls belie the almost

© The Author(s) 2019   81
D. Reay, *Climate-Smart Food*,
https://doi.org/10.1007/978-3-030-18206-9_7

complete lack of genetic diversity in the global banana crop itself. Bananas were first domesticated several thousand years ago in Southeast Asia, with the ancestors of the bananas we eat today growing wild in Papua New Guinea and Indonesia. By the time the first commercial shipment of refrigerated bananas reached Britain in 1902 [1], they were being grown in tropical and subtropical regions all around the world.

Banana plants like warm and wet conditions, along with fertile soils. They grow best in the tropics, with an average temperature in the high-20s Celsius, and can be found in plantations in a wide band between 30 degrees north and south of the equator. Commercial plants are usually grown to just a few metres tall and are replanted every six years or so using the bulbous shallow-rooted rhizomes (called corms) that the adult plants produce [2]. Today some 130 countries grow bananas, with the 120 million tonnes they produce each year mainly being grown by smallholders for home consumption or sale at local markets. Bananas have become one of the most important crops in the world and are a vital part of human diets in many regions—the inhabitants of New Guinea are estimated to consume an impressive 200 kilograms of bananas per year each (in Europe and North America we average a relatively paltry 15 kilograms per person). Globally, it is India and China that dominate world supplies, along with high production in Central and South America [3] (Fig. 7.1).

Across the scores of nations and myriad plantations that grow bananas there are many hundreds of 'cultivars' too—selectively bred varieties that have a wide array of shapes, colours, size and tastes. These cultivars include the sweet, yellow-skinned inhabitants of our Western lunch boxes and the more starchy plantains widely used for cooking. They are wonder fruits in terms of the nutritional punch they pack, being rich in potassium as well as in vitamins A and C [5].

Almost all are descended from just two wild species and this lack of underlying genetic diversity is a serious risk factor when it comes to disease. In the lean post-war years of the 1950s bananas gradually began to return to the ration-hit grocery shelves. Then disaster overtook the world's banana growers, in the form of a fungus. The most popular variety at the time was *Gros Michel* (Big Mike) and one plantation after the next succumbed as a *Fusarium* fungus causing so-called Panama Disease ran rampant. This fungus attacks roots and eventually kills the whole plant. Its post-war outbreak cost over $2 billion [6]. With no effective treatment, farmers were forced to switch to the more resistant variety—Cavendish—that we mainly eat today. The fungus didn't disappear though. In the early

Banana production, 2014
Annual banana production, measured in tonnes per year.

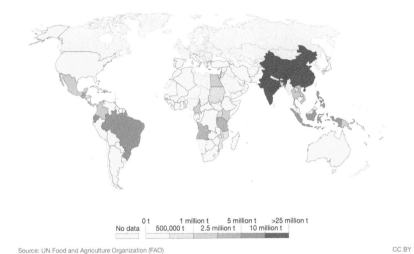

| No data | 0 t | 500,000 t | 1 million t | 2.5 million t | 5 million t | 10 million t | >25 million t |

Source: UN Food and Agriculture Organization (FAO)   CC BY

Fig. 7.1 Global banana production in 2014 by country of origin (Source: Hannah Ritchie, Our World in Data) [4]. Available at: https://ourworldindata. org/grapher/banana-production

1990s a new strain of it called Tropical Race 4 emerged that could attack the Cavendish banana plants too. Tropical Race 4 has already spread across Southeast Asia and into East Africa and the Middle East. It can wipe out whole plantations and as yet no resistant banana variety that can easily replace the under-siege Cavendish has been found [7]. Adding fuel to the banana disease fire comes climate change. Through direct impacts like storms, drought and heat, to indirect threats like accelerating disease spread, it is an unwelcome complication for growers already facing a highly uncertain future.

Our own break time banana here in Scotland originated in the balmier climes of the Dominican Republic—a sweet-spot on the global growing map and currently the world's biggest supplier of organic bananas [8]. Produced by the thousand or so farmers on the island, almost all of the banana exports from the Dominican Republic end up in Europe.

Along with the rest of the roughly 800,000 tonnes of bananas imported to the UK each year, my own banana has travelled a very long distance, has been refrigerated along the way, and so carries with it a substantial carbon footprint. However, as far as bananas go, it is relatively low carbon. The average kilogram of bananas results in the emission of 0.5–1.3 kilograms of greenhouse gas—or 100–200 grams per banana.

In the tropical plantations, use of nitrogen fertilisers (causing emissions of nitrous oxide from the soil) and the energy used (for machinery and to make fertilisers and pesticides), are the main sources of on-farm emissions. More are racked up in harvesting, processing (fungicide sprays and chlorine dips to remove the sticky latex that oozes from cut stems), and packaging. For most bananas that end up on distant foreign shores it is then transport that adds a big extra dollop of carbon [9].

The journey from plantation to port tends to be short and by diesel truck. The bananas are then loaded onto refrigerated ships—called reefers—or container ships at coastal towns like Manzanillo to begin their two-week journey across nearly 4,000 miles of ocean. These ships are powered by heavy fuel oil, and if the ships are inefficient and don't have a cargo to carry on the return trip, their emissions can account for up to two-thirds of the total banana footprint [10]. Once imported, the boxes of bananas are taken to regional centres for ripening and then distribution to shops and supermarkets. Once we buy them, a final race against time to avoid an even larger climate penalty begins.

Few banana lovers have avoided the pain of a seemingly hard and green bunch one day becoming blackened mush the next. There is the slimy horror of a forgotten banana in the bottom of a school bag or the pungent surprise of a benthic banana layer in the fruit bowl.

Keeping bananas wrapped and the bunches separated can apparently slow how fast they ripen—ripe bananas emit ethylene gas which will accelerate ripening in any nearby bananas if they aren't protected. Refrigeration can prolong their lifespan but, as has caused plenty of arguments in our house, everything else in the fridge risks taking on a banana-ey taste.

Having travelled a long way and already with a weighty carbon footprint, the wastage of bananas by consumers contributes a major part to their life-cycle climate change impact. It also represents the prime target for most of us to reduce it. Avoidable wastage (so not including the banana skins) in the UK alone tops 65,000 tonnes a year—over a million bananas every day [11]. Most of this is simply through them not being used in time and means 30,000–65,000 tonnes of unnecessary greenhouse gas is emitted annually. Each banana we save from the bin—by keeping an eye on their ripeness and avoiding overbuying—helps to avoid the growing, processing and transport required to deliver its replacement. All those parts of the banana's life cycle can be targeted directly too.

On the farm, a reduction in fertiliser and pesticide use—as achieved with our own organic bananas in the Dominican Republic—means a smaller footprint compared to their conventionally produced cousins [12]. For transport over land, lower driving speeds, alternative fuels and improved logistics can lower emissions, while in the ripening and distribution centres more energy efficient air-conditioning and leak-free refrigeration will cut both costs and carbon. For the biggest part of banana's carbon footprint—shipping—a switch away from traditional reefer ships to larger, more efficient, refrigerated container ships can slice over a third off transport emissions. Coupled with lower cruising speeds, computer-controlled storage, and low-carbon fuels and refrigerants, addressing shipping emissions is a prime target [9]. Globally, shipping now accounts for 2 per cent of all greenhouse gas emissions and pressure is mounting on the wider industry to curtail its use of fossil fuels [13].

* * *

With a global banana industry under siege from disease, and climate change adding an unwelcome risk to the mix, climate-smart approaches have become more important than ever [14]. Banana plants like it warm and wet, and some plantations are already on the margins of the viable climate envelope. As each crop cycle takes 12 months or more, they are also exposed to the full year's worth of extreme weather events. Drought is the climate risk that most commonly concerns banana growers [15]. Rainy seasons are set to become less reliable and farmers in the drier sub-tropical zones face a damaging combination of reduced rainfall and high temperatures. By the 2050s, climate change—mainly via drought impacts—is expected to have wiped over a million square kilometres off the viable banana-growing map in the sub-tropics [16]. In the second half of the century extreme temperatures are also predicted to take a toll, with plantations in some areas of Argentina and India having to endure temperatures in excess of 40 degrees Celsius through the summer season. However, this shifting climate envelope for bananas should mean a major boost in production in some parts of the tropics, and a northwards extension of the growing range in Europe, Asia and North America. Over 5 million square kilometres is expected to shift out of the too-cold-for-bananas climate zone by 2070 and, overall, the global growing area for bananas is expected to increase due to climate change in the twenty-first century.

Our growers in the Dominican Republic can expect a mixed bag of long-term climate impacts. The news on temperature looks okay, with a predicted rise of 2 to 3 degrees Celsius by the 2070s [17], meaning new summer highs of up to 30 degrees Celsius and warmer winter minimums of 20 degrees Celsius. The banana plants should, therefore, still have a comfortable growing temperature throughout the year. Rainfall is more of a concern. As climate change intensifies so the amount of wet season rain in the Dominican Republic is set to fall and the dry season to extend [16]. Without irrigation this may raise the spectre of drought risks in the long term, but for the farmers of the Dominican Republic there's already a major weather threat to contend with: hurricanes.

The Atlantic hurricane season of 2017 was devastating. It encompassed 17 named storms of which Hurricanes Harvey, Maria and Irma were the most violent. Tropical storm Maria emerged on the 16th of September and was to be the deadliest storm of them all. Picking up energy over an especially warm ocean, the storm grew rapidly, and by the 18th of September it had become a Category 5. As it raged across the Caribbean it claimed the lives of over 140 people and left a trail of destroyed homes, infrastructure and livelihoods.

Maria passed to the north and east of the Dominican Republic on the 21st of September 2017, bringing damaging high winds and depositing huge amounts of rain in just a few hours. Around 5,000 hectares of banana plantations were flooded. Earlier high winds during storm Irma had already bent and broken many banana plants, now, with waterlogged soils too, many plants suffered irreparable damage [18]. Some growers saw 80 per cent of their plantations destroyed in that ferocious 2017 storm season, the wet conditions and damaged plants proving a ripe breeding ground for fungal diseases too [19].

Whether climate change will alter storm and hurricane risks in the Dominican Republic, and more widely, is still uncertain. Warmer seas have the potential to increase the energy of storms and result in more powerful hurricanes, but changing wind shear (the difference in speed and direction of winds in the upper and lower levels of the atmosphere) may actually reduce their number [20]. Whatever the net effect of these two opposing forces, it is likely that the kind of extreme rainfall and flooding risks posed by the storm season of 2017 will be even greater in the coming decades.

*   *   *

The huge importance of bananas to the livelihoods of so many people around the world has already spurred climate-smart action. Improved knowledge and access to training has been central to these efforts. In Costa Rica, for instance, the banana industry supports over 100 thousand families [21] and a joint effort by the government and development agencies has now produced practical guidance for growers and suppliers. Its aim is to embed greater climate resilience while simultaneously boosting production and cutting emissions [14]. Likewise, the large banana industries in Ecuador [22] and the Dominican Republic are benefiting from increased research and sharing of good practice, with the Dominican Republic seeing more and more assessment of how its organic production methods could help deliver greater protection from drought and pest risks [8].

Water is a concern for many growers. Keeping their thirsty plants well-watered throughout the year often requires the use of irrigation, but changing rainfall patterns, and increasingly unreliable supplies from glacial melt in areas like Ecuador, mean improved water management is a central part of the climate-smart response. The use of drip irrigation methods [23] and more rainwater collection and storage [24] can help buffer the effects of drought. If used alongside good weather forecast information and crop water demand advice, the water supply can be much better matched to the needs of the plant. As banana plants have shallow roots, mulching their soils with clippings or planting with low-growing cover crops can also provide a shaded understory that helps to maintain soil moisture even when temperatures soar [9]. Such mulching and use of cover crops has the added benefits of suppressing weeds, reducing soil erosion, and cutting run-off of pesticides and herbicides during times of very heavy rainfall [25].

With disease resistance being the urgent focus of most banana-breeding programmes around the world, research on more drought-resistant plants has so far been limited. Some varieties do show more drought tolerant traits though and, for those farmers facing the biggest drought risks or having the least ability to adapt through irrigation, development and supply of new hybrids able to withstand drier conditions could be vital [5].

Like too little water, too much water can also pose a serious risk for bananas in the form of disease. Here, fungal diseases like the dreaded Black Sigatoka become a real threat [26]. Since 2008 this devastating fungus has spread through all the main banana-producing areas of the Caribbean—in Guyana it annihilated banana production in just three years

[27]. Black Sigatoka thrives in wet, warm conditions, with heavy rain-storms and flooding allowing it multiply and spread quickly. Though fungicide treatment can help control it, these fungicides are expensive. Using expert advice and support, many farmers have now responded by boosting overall plant health and disease resilience through better fertiliser use, removal of infected plants, and improved soil drainage [28]. Careful timing and application of the expensive fungicides has also allowed more farmers to use them, with the added bonus of reducing water pollution risks during heavy rain. Production in Guyana is already recovering, and a region-wide strategy is now planned where best practice is shared across all nations. This plan includes providing disease-free rhizomes to growers, developing new disease-resistant varieties, and building public awareness and the capacity to respond to future outbreaks [27].

For the powerful storms and hurricanes that threaten the Caribbean each year, early warning systems and recovery plans may help save bananas as well as many lives. The banana plants themselves grow from a large rhizome, so even if a storm damages leaves and stems they can often regenerate if given the chance. Where wind damage destroys large numbers of plants, as can happen with hurricanes, having robust systems for recovery in place—like rapid infrastructure repair, disease-free rhizome and equipment re-supply—can help limit the overall impacts and get production up and running quickly [29].

In the face of more extreme weather, and with diseases like fusarium wilt and Black Sigatoka threatening to wipe out their livelihoods, many banana growers are turning to diversification—planting other crops alongside their bananas—as a way to give more resilience. One of the most successful of these banana-growing marriages is with coffee.

Coffee grows in similar conditions to bananas and is itself at major risk from climate change impacts, such as through higher temperatures. Growing it alongside bananas means the banana leaves provide much-needed shade during the hottest parts of the day and often means a more consistent coffee crop. The residues from the banana plants also provide a useful mulch for the coffee plants, helping to boost fertility and carbon storage in the soils, retain soil moisture and supress weeds. This winning combination provides a valuable insurance policy for farmers in the event of disease attack and loss of one of the crops.

The nutrient and water needs of the two crops do have to be well balanced—there is a risk that the coffee will outcompete the banana plants if both aren't supplied with enough fertiliser and water, or that planting the bananas too close together will shade the coffee too much. Such banana-coffee intercropping strategies have been known about for decades and have already been successful on farms across Latin America, Africa and Asia [30]. Research on banana-coffee combinations in Uganda has reported increases in yield values and farmer income of over 50 per cent, alongside improved coffee quality, resilience to extreme weather, and more carbon storage in the plants and the soils beneath them.

For banana growers around the world, climate change represents a real risk multiplier for the major threats already posed by disease and severe weather events. Climate-smart approaches, local capacity-building, and enhanced regional systems that allow sharing of best practice and support can all help ensure a banana-shaped future of our break time.

REFERENCES

1. Mckie, R. We're all going bananas. *The Observer*. https://www.theguardian.com/observer/focus/story/0,6903,746576,00.html. (2002).
2. FAO. *Banana*. http://www.fao.org/land-water/databases-and-software/crop-information/banana/en/ (2018).
3. FAO. *Banana Facts and Figures*. Food and Agriculture Organization of the United Nations. http://www.fao.org/economic/est/est-commodities/bananas/bananafacts/en/#.XOqq0y2ZPUo (2018).
4. Ritchie, H. Global banana production, 2014. *Ourworldindata.org*. https://ourworldindata.org/grapher/banana-production (2018).
5. Ravi, I., Uma, S., Vaganan, M. M. & Mustaffa, M. M. Phenotyping bananas for drought resistance. *Front. Physiol.* **4**, 9 (2013).
6. Fusariumwilt.org. *Panama Disease*. https://fusariumwilt.org/index.php/en/about-fusarium-wilt/ (2018).
7. FAO. *Banana Fusarium Wilt Disease*. http://www.fao.org/food-chain-crisis/how-we-work/plant-protection/banana-fusarium-wilt/en/ (2018).
8. FAO. *Organic Banana Production in the Dominican Republic*. Food and Agriculture Organization of the United Nations. http://www.fao.org/world-banana-forum/projects/good-practices/organic-production-dominican-republic/en/#.XOqrCC2ZPUo (2018).
9. FAO. *Carbon Footprint of the Banana Supply Chain*. Food and Agriculture Organization of the United Nations. http://www.fao.org/3/a-i6842e.pdf (2017).
10. Iriarte, A., Almeida, M. G. & Villalobos, P. Carbon footprint of premium quality export bananas: Case study in Ecuador, the world's largest exporter. *Sci. Total Environ.* **472**, 1082–1088 (2014).

11. WRAP. Household food and drink waste in the United Kingdom 2012. *Waste and Resource Action Programme*. http://www.wrap.org.uk/sites/files/wrap/hhfdw-2012-main.pdf.pdf. (2013).

12. Roibás, L., Elbehri, A. & Hospido, A. Carbon footprint along the Ecuadorian banana supply chain: Methodological improvements and calculation tool. *J. Clean. Prod.* **112**, 2441–2451 (2016).

13. Gabbattis, J. Carbon emissions from global shipping to be halved by 2050, says IMO. *The Independent*. https://www.independent.co.uk/environment/ships-emissions-carbon-dioxide-pollution-shipping-imo-climate-change-a8303161.html (2018).

14. FAO. *Reducing Carbon and Water Footprints in Banana Plantations*. Food and Agriculture Organization of the United Nations. http://www.fao.org/world-banana-forum/projects/reducing-carbon-and-water-footprints-in-banana-plantations/en/ (2018).

15. Calberto, G., Blake, D., Staver, C., Carvajal, M. & Brown, D. *X International Symposium on Banana: ISHS-ProMusa Symposium on Agroecological Approaches to Promote Innovative Banana 1196* 179–186.

16. Calberto, G., Staver, C. & Siles, P. An assessment of global banana production and suitability under climate change scenarios. *Climate Change and Food Systems: Global Assessments and Implications for Food Security and Trade* (Food Agriculture Organization of the United Nations (FAO), Rome, 2015).

17. Carbonbrief. Mapped: How every part of the world has warmed—And could continue to warm. *Carbonbrief.org*. https://www.carbonbrief.org/mapped-how-every-part-of-the-world-has-warmed-and-could-continue-to-warm (2018).

18. producebusinessuk.com. *A Quarter of Farms Flooded for Major Banana Exporter*. https://www.producebusinessuk.com/supply/stories/2017/09/27/a-quarter-of-farms-flooded-for-major-banana-exporter (2017).

19. freshplaza.com. *Dominican Republic: "We Have Losses of up to 50 Percent"*. https://www.freshplaza.com/article/185965/Dominican-Republic-We-have-losses-of-up-to-50-percent/ (2017).

20. GFDL. *Large-scale Climate Projections and Hurricanes*. Geophysical Fluid Dynamics Laboratory, NOAA, USA. https://www.gfdl.noaa.gov/global-warming-and-21st-century-hurricanes/ (2018).

21. Alvarado, L. Costa Rica leads the way in carbon neutral banana production. *Costa Rica Star*. https://news.co.cr/carbon-neutral-banana-production-costa-rica/70539/ (2018).

22. Elbehri, A. *et al. Ecuador's Banana Sector Under Climate Change*. Food and Agriculture Organization of the United Nations. http://www.fao.org/3/a-i5697e.pdf (2016).

23. Silva, A. J. P. d., Coelho, E. F., Miranda, J. H. d. & Workman, S. R. Estimating water application efficiency for drip irrigation emitter patterns on banana. *Pesq. Agropec. Bras.* **44**, 730–737 (2009).

24. FAO. *Feasibility Study of Rainwater Harvesting for Agriculture in the Caribbean Subregion.* Food and Agriculture Organization of the United Nations. http://www.fao.org/3/a-bq747e.pdf (2014).
25. Bellamy, A. S. Banana production systems: Identification of alternative systems for more sustainable production. *Ambio* **42**, 334–343 (2013).
26. freshfruitportal.com. *Black Sigatoka Fungus Hits Dominican Republic's Bananas.* https://www.freshfruitportal.com/news/2011/11/21/black-sigatoka-fungus-hits-dominican-republics-bananas/ (2011).
27. FAO. *Battling Black Sigatoka Disease in the Banana Industry.* Food and Agriculture Organization of the United Nations. http://www.fao.org/3/a-as087e.pdf (2013).
28. Ploetz, R. Black Sigatoka of banana: The most important disease of a most important fruit. *APS Feat* **2**, 126. https://doi.org/10.1094/PHI-I-2001-0126-02 (2001).
29. FAO. *Post Disaster Damage, Loss and Needs Assessment in Agriculture.* Food and Agriculture Organization of the United Nations. http://www.fao.org/docrep/015/an544e/an544e00.pdf (2012).
30. Asten, P. V. *et al. Coffee–Banana Intercropping: Implementation Guidance for Policymakers and Investors.* Global Alliance for Climate-Smart Agriculture. https://cgspace.cgiar.org/bitstream/handle/10568/69017/CCAFSpbCoffee-Banana.pdf (2015).

CHAPTER 8

Climate-Smart Coffee

Abstract Coffee is grown widely along a tropical bean belt stretching across Central and South America, Africa and Asia. Brazil alone produces over two million tonnes a year and global production now tops ten million tonnes annually—over two billion cups of the stuff every day. Coffee's life-cycle carbon footprint ranges from around 70 grams per cup for instant to as much as 150 grams per cup for filter coffee. Major pests and diseases like leaf rust and the coffee berry borer are predicted to become even more prevalent under a future climate. In the highlands of Ethiopia—coffee's birthplace—warmer and wetter conditions in the future may allow fungal diseases to spread to higher altitudes and so threaten areas that are so far fungus free. Moving farms uphill, using shade and irrigation, and potentially switching to new hybrid varieties can all boost resilience and help reduce emissions. Access to training, advice and technology remains a major barrier to this for many coffee farmers.

Keywords Shade coffee • Agroforestry • Ethiopia • Arabica • Robusta • Hybrids • Rift Valley • Coffee berry borer • Diversification

Whether early morning rocket fuel, work meeting mainstay or after-dinner indulgence, coffee is a beloved drink for many millions of us around the world. Our long-standing love affair with it can be traced back to ninth-century Abyssinia (modern Ethiopia) and the province of Kaffa in the

© The Author(s) 2019   93
D. Reay, *Climate-Smart Food*,
https://doi.org/10.1007/978-3-030-18206-9_8

southwestern highlands that gave this caffeine-rich infusion its name [1]. Legend has it that a goatherd named Kaldi first discovered wild coffee plants. One hot afternoon in the hills, Kaldi's goats started behaving very oddly. They jumped and skipped, sprinted and surged up the slopes ahead of him. Kaldi saw that the nearest goats were busy eating some sort of red berries from scattered, low-growing bushes. He collected a handful of the berries and tasted them—they were bitter and had large seeds. Kaldi spat them out, but already a feeling of elation was spreading through him. His heart rate quickened, new energy surged through his weary legs. These unappetising berries were the reason for the prancing and leaping of his goats, now he felt like prancing and leaping too.

Collecting more of the precious fruit, Kaldi headed down into the valley to tell his story and present his super-charged berries to the abbot of the local monastery. The abbot was wary. Was this the devil's work? Some sort of poison perhaps? Kaldi looked okay, although he couldn't keep still for long. The abbot had his monks make a brew of the crushed berries and asked his novices to try some. First one, then two, then all of them grinned as their energy levels surged. The abbot and his monks soon found that chewing on just a handful of Kaldi's berries could keep them awake and alert throughout their long nights of prayer. Word quickly spread about the wonder drink from Kaffa. Coffee was born.

By the fifteenth century coffee plants were being cultivated on the Arabian Peninsula, and by the seventeenth century it was being drunk in the chattering coffee houses of London, Paris and Rome. Today coffee is grown widely along a tropical bean belt stretching across Central and South America, Africa and Asia. Our appetite for it is huge. Brazil alone produces over two million tonnes a year and global production now tops ten million tonnes annually (Fig. 8.1). That's over two billion cups of the stuff every day [2].

For dozens of developing countries, coffee has become the aromatic keystone of their economies. In the nations of Burundi and Uganda for example, it accounts for over half of all foreign currency earnings. Worldwide, over 100 million people rely on it for their livelihoods [3], with the US and Europe being the biggest coffee importers [4]. Global trade is now worth around $10 billion each year—second only to petroleum in the rollercoaster ride of international commodity prices [5].

Around two-thirds of the coffee we drink is called Arabica and is produced from roasting the twinned large seeds (coffee beans) encased in the berries of the *Coffea arabica* plant. Most of the rest is known as Robusta

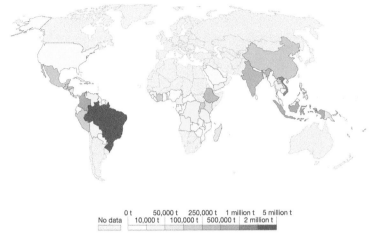

Coffee bean production, 2014
Annual coffee bean production, measured in tonnes per year.

Fig. 8.1 Global coffee bean production in 2014 by country of origin (Source: Hannah Ritchie, Our World in Data) [6]. Available at: https://ourworldindata.org/grapher/coffee-bean-production

coffee and comes from the *Coffea canephora* plant—a close cousin of Arabica first reported in the Democratic Republic of Congo. Robusta, as its name suggests, is easier to cultivate, can have bigger yields, and is more resistant to disease thanks to its high concentrations of caffeine and antioxidants. It also tends to have more of a bitter taste than Arabica and so is most commonly found in lower cost instant coffees. Exploiting the differences between the delicate premium bean-producing Arabica and the tougher Robusta may hold the key to ensuring a climate-smart future for our daily fix.

Both plants enjoy a tropical climate, rich soils and plenty of rainfall (around 1,800 millimetres a year). They produce their first crop of rich red berries 2 to 3 years after planting and can go on for a further 30 years as productive plants [5]. Arabica is the fussier of the two, requiring a distinct rainy and dry season and a year-round temperature range of 15 to 24 degrees Celsius. Above and below this range growth rates start to plummet, with frost damage likely as the plants approach zero. Robusta is even more sensitive to the cold, but seems to thrive at higher temperatures and can cope without there being defined wet and dry seasons.

Our own daily coffee-high in West Lothian comes courtesy of a blend of ground Arabica filter coffee. It includes beans sourced from South America and Africa, with Ethiopia (coffee's birthplace) being the biggest African producer of the particular high strength rocket fuel we have come to love. Ethiopia has an ideal climate for coffee growing, its swathes of high-altitude land mean it can produce some of the best-quality beans in the world, and in large quantities. Coffee provides a livelihood for some 15 million Ethiopian smallholders and their families, almost a fifth of the population. The evergreen forests, high moisture and cooler mountain temperatures of the highlands allow for slower-growing coffee plants that make for premium coffee beans.

Ethiopian farmers mostly hand pick their ripe red coffee berries. For the top-priced beans only the reddest, ripest berries are picked, while for lower grade coffees all the berries are picked in one go. Once picked, the clock starts ticking on getting the crop processed—growers have just 12 hours to get their precious harvest to a washing station (called a pulpery) before it starts to degrade [7]. The pulperies collect ripe berries from all the farmers in the vicinity and soak them in water for 2 to 3 days. This allows removal of the pulpy outer layers to leave the twin 'beans' encased in a slippery skin. These seeds are then dried and the inner skin removed to produce green coffee beans ready to be shipped to the big auction houses of Addis Ababa and Dire Dawa. Export to the consumers of the world is usually via container ship, the precious cargoes then being delivered to roasting plants where the distinctive flavours and aromas of the coffee are brought to the fore—the longer the roasting, the darker the coffee.

The final leg in the life of our roasted beans depends on their destination. Some are shipped as whole beans to be ground to order in one of the countless thousands of coffee shops that now inhabit our streets and shopping centres—in the UK alone we have over 20,000, such coffee shops serving us more than two billion cups of coffee a year [8]. For filter coffees, the beans are ground, bagged and sent to retailers, while for instant coffee the ground coffee is made into a series of stronger and stronger brews. The powerful liquor is then either frozen to minus 40 degrees Celsius in a vacuum (freeze dried) or dried by spraying droplets through a stream of hot air (spray dried).

All of these stages play a role in the total carbon footprint of our coffee, with the growing stage and how we as consumers then choose to prepare and drink it being the prime ones. On the plantations, it is again the use of nitrogen fertilisers that tend to dominate coffee's climate change impact. More greenhouse gas comes from land use change and soil disturbance, and from the energy used in irrigation and pesticide production. Because filter coffee uses more coffee per cup (around 9 grams) than that of the instant types (around 2 grams) its carbon footprint from the growing phase is bigger: over 20 grams of greenhouse gas emissions per cup for filter coffee, versus under 10 grams per cup for instant.

Processing and packaging tips the climate scales in the other direction, with instant coffees needing more packaging and so generating higher emissions there. The packaging for the espresso capsule or pod-type coffees that have become the latest dust-gathering gadget in many kitchens has an especially big carbon footprint, being almost ten times that for the filter coffee and representing the single biggest part of the capsules coffee's climate impact [9]. The capsules themselves can also be hard to recycle [10]. Transport and distribution then add their own top-ups to each coffee's carbon footprint before we then get to decide what the final climate bill will be.

Coffee's life-cycle carbon footprint ranges from around 70 grams per cup for instant to as much as 150 grams per cup for filter coffee. Boiling water and the energy used for making and washing of the coffee cups is a major player, racking up almost half of the total emissions. As such, we have considerable power to make our daily pick-me-up lower carbon. Boiling only the water required [11] will slash these 'consumption' phase emissions. Likewise, avoiding waste—measuring out the perfect amount of grounds and reheating lukewarm drinks—and using a reusable coffee cup can ease the total emissions over time [12, 13]. An estimated 2.5 billion disposable coffee cups are used in the UK each year, resulting in 30,000 tonnes of rubbish [14]. Only a fraction of this is currently recyclable.

Since Kaldi's goats provided the first documented caffeine buzz over a thousand years ago, coffee drinking has become a major player in the global climate impact of our food and drink. The two billion or so cups of coffee drunk each day mean our worldwide coffee addiction is now responsible for around 70 million tonnes of emissions each year. Moving from our households back along the lengthy coffee supply chains are numerous other opportunities to reduce its climate impact, from lower-emission ships and trucks in its transport, to the use of renewable energy in its processing and roasting phases.

A climate-smart approach needs all this and more—providing a lower carbon cup of coffee, that is resilient and that brings its growers secure livelihoods. For the millions of smallholders who grow coffee in Ethiopia the battle to achieve this is already raging.

*   *   *

Our love for Arabica coffee beans combined with its pernickety growing requirements and a changing climate make for an expensive cocktail. While consumption has doubled in the last 35 years, global production of these premium beans has been hobbled by severe weather and disease outbreaks.

In Ethiopia the coffee-growing heartlands are to be found in the Rift Valley and in the cool tropical forest areas to the west and east of this. At altitudes of 1,000–2,000 metres, these high forests provide the ideal growing conditions for Arabica coffee plants—the shade they provide protects the coffee plants from extremes in sunlight and temperature while also maintaining humidity. The right amount of rainfall, and at the right time, is crucial. But annual rainfall has been decreasing in south-west Ethiopia since the 1950s and the timing and length of the rainy season has become more and more unpredictable. In 2015 a severe drought hit the Harar coffee zone in the eastern highlands. Plants began to wilt and their leaves to curl. As the drought progressed the leaves fell and the few beans that did grow became deformed. By the end of the dry season in early April large numbers of coffee plants, covering many hundreds of hectares, lay dead [15]. In the following year much of the country was hit by an even more severe drought, the worst in 50 years [16].

Rising temperatures make such drought impacts more likely—drying the soils faster, putting extra stress on the plants and reducing the quality of the beans. By the middle of this century average temperatures are set to have increased by over 2 degrees Celsius across much of Ethiopia, with the end of the century seeing in excess of 4 degrees Celsius of warming [17]. At the same time average rainfall across Ethiopia is actually likely to increase, but with a greater risk of extreme rainfall events (causing soil erosion and flooding) and more unpredictability in the timing of wet and dry seasons [15].

Coffee growers in Ethiopia and around the world must face the spectre of pest and disease attack too. Some of the biggest current threats, such as Leaf Rust and the coffee berry borer, are predicted to become even more prevalent under a future climate. Leaf Rust is a fungal disease that is endemic to Ethiopia that has now spread to all coffee-growing regions of the world. Across the million or so hectares of plantations in Colombia an outbreak of Leaf Rust was blamed for the loss of more than a third of production between 2008 and 2011 [18]. In the highlands of Ethiopia, warmer and wetter conditions in the future may allow this fungus to spread to higher altitudes and so threaten areas that have so far been fungus free.

The coffee berry borer appears to be on the march too. A decade ago it was never seen in plantations above about 1,500 metres, yet warming is today allowing it to spread to ever-higher altitudes and with further warming it has the potential to affect over three-quarters of plantations [19]. This costly insect is the most important coffee pest worldwide, causing damage estimated at over $500 million a year and so putting at risk the more than 25 million livelihoods that depend on the industry [20].

The bitter brew of higher temperatures, more unpredictable rainfall, and increased attack from pests and disease means that over half of the current coffee-growing areas in Ethiopia risk being squeezed out of existence by climate change [21].

To escape drought, heat and pests impacts at lower altitudes, more and more growers are expanding their coffee farming higher up the mountain slopes. The Ethiopian government is helping to support this kind of active adaptation, with new plantations being encouraged at heights of over 3,000 metres (a kilometre higher than the norm) [22]. In fact, as Ethiopia's climate changes, some areas are likely to see coffee-growing conditions becoming much better even while others falter.

In the Rift Valley and to its southwest, the area suitable for Arabica coffee is expected to expand through until the middle of the century. This is especially true of higher ground, where warmer conditions will allow successful production in areas that were previously too cold. For a climate-smart response, good access to such locally-specific information is vital. Over in the southeast and eastern highlands, in already drought-hit areas like Harar, the situation looks more worrying. Major declines in the area suitable for coffee are predicted, with complete loss of all viable areas possible in the second half of the century. For Ethiopia's millions of smallholder coffee farmers the future lies somewhere between an extreme of no action—where huge swathes of production are lost, to one of proactive strategies and migration in which coffee farming could boom to four times its current size [23].

* * *

The migration of coffee growing to areas more suitable in a future Ethiopian climate certainly makes sense in terms of increased resilience. To achieve this in a climate-smart way—accounting for its impacts on greenhouse gas emissions and on productivity too—will require careful planning and assessment of the new areas. The specific soil, drainage and shading needs of the Arabica plants need to be met alongside their climate requirements. In some areas there are risks of conflict with other land uses and, because the coffee plants often grow best in shade, planting of companion shade trees in currently unforested areas will be needed. Where successful, such tree planting offers a good opportunity to combine coffee migration with increased carbon uptake.

This practice of shade-coffee is already widely practised around the world including, as we have seen, in combination with bananas (Chap. 7) [19]. For those coffee farmers in Ethiopia already battling the impacts of drought, heat and disease it may boost resilience and extend the viable lifetime of their plantations. The shade trees can cut the temperatures under the canopy by around 4 degrees Celsius compared to unshaded areas, and can serve to reduce damage from intense rainfall or desiccating winds [15].

Down on the ground, mulching of soils to reduce water loss and supress weeds can be effective, while irrigation in drought-prone areas also has the potential to increase resilience in a future climate. The use of irrigation by smallholders remains limited though, and much greater awareness and financial support is required before it reaches its full potential. An effective strategy that is already commonly used is that of terracing (creating flat areas of land bounded by embankments). Such terracing allows farmers to better control drainage and soil moisture, and so to limit drought risks.

For the coffee plants themselves, a climate-smart approach includes good matching of their fertiliser and water needs with what is supplied. As nitrous oxide emissions from nitrogen fertiliser use is one of the biggest components of coffee's carbon footprint, limiting fertiliser inputs to only what is needed for optimal growth benefits both the farmer's profits and the climate [24].

For pesticide use, too, the sharing of good practice and provision of more sustainable alternatives can improve productivity and reduce pollution risks. Many coffee growers rely on the highly hazardous insecticide Endosulfan to fight the ravages of the coffee berry borer. This toxic chemical is, however, a persistent organochlorine, meaning that once it enters the food chain it can accumulate and become toxic to animals and humans—for smallholder farmers in the developing world pesticide exposure has become the leading occupational hazard [25]. As an alternative to Endosulfan, coffee growers in Colombia, Nicaragua and El Salvador have successfully used improved field hygiene methods (such as removing damaged or fallen berries), trapping of the female borers with alcohol attractants, and promotion of biological controls like the insect-killing fungus *Beauveria bassiana* [26].

In some areas no amount of improved care will save the premium Arabica coffee plant and instead it will have to give way to its more heat-tolerant cousin 'Robusta'. The price per kilo of coffee produced may fall, but if total production is increased and gives a more reliable income then smallholder coffee farmers in increasingly marginal growing areas can be better protected.

Beyond switching to Robusta, alternative coffee plant varieties—such as new hybrids of Arabica and Robusta—arguably offer the greatest potential for more drought, pest and disease resistance. Work is underway to establish which existing varieties show the most genetic diversity, with the hope that from these will come hybrid plants better suited to specific areas and better adapted to what climate change will throw at them [27]. The so-far-identified 124 wild coffee species of the world are vital to these efforts, yet 60 per cent of these are now on the edge of extinction due to habitat loss [28].

Here, and across the range of climate-smart approaches for coffee, good research, information and support are vital. In Kenya, Ghana and Zambia, an expanding Pest Risk Information Service is combining weather data, computer models and local crop monitoring to give farmers early warning on pest and disease attacks [29]. Such extension services, finance and opportunities to share best practice between farmers could make all the difference in a changing climate. In addition to improving information and advice on growing practices and migration planning, these farmer-level support systems can open up opportunities to new markets, better prices, and core resilience services like crop insurance [30].

The world's coffee growers are in the midst of a literal uphill struggle to meet soaring demand in the face of increasing pressures from climate change and disease. As a global commodity its carbon footprint is large and the number of livelihoods that depend on it huge. Climate-smart approaches today, from bright berried bushes in Ethiopia to our morning caffeine fix in Scotland, can help to ensure coffee will still be keeping our great grandchildren awake tomorrow.

REFERENCES

1. Nzegwu, N. *Ethiopia: The Origin of Coffee*. https://www.africaresource.com/house/news/our-announcements/21-the-history-of-coffee (1996).
2. USDA. *Coffee: World Markets and Trade*. United States Department of Agriculture. https://apps.fas.usda.gov/psdonline/circulars/coffee.pdf (2018).
3. Bunn, C., Läderach, P., Rivera, O. O. & Kirschke, D. A bitter cup: Climate change profile of global production of Arabica and Robusta coffee. *Clim. Chang.* **129**, 89–101 (2015).
4. Fairtrade. *About Coffee*. https://www.fairtrade.org.uk/en/farmers-and-workers/coffee/about-coffee (2018).
5. Nair, K. P. *The Agronomy and Economy of Important Tree Crops of the Developing World* (Elsevier, 2010).
6. Ritchie, H. Global cocoa bean production, 2014. *Ourworldindata.org*. https://ourworldindata.org/grapher/cocoa-bean-production (2018).
7. Mutua, J. *Post Harvest Handling and Processing of Coffee in African Countries*. Food and Agriculture Organization of the United Nations. http://www.fao.org/docrep/003/x6939e/X6939e12.htm (2000).
8. Hooker, L. Is the UK reaching coffee shop saturation point? *BBC Online*. https://www.bbc.co.uk/news/business-41251451 (2017).
9. Humbert, S., Loerincik, Y., Rossi, V., Margni, M. & Jolliet, O. Life cycle assessment of spray dried soluble coffee and comparison with alternatives (drip filter and capsule espresso). *J. Clean. Prod.* **17**, 1351–1358 (2009).
10. BBC. *Is There a Serious Problem with Coffee Capsules?* https://www.bbc.co.uk/news/magazine-35605927 (2016).
11. Aldred, J. Tread lightly: Keep your kettle in check. *The Guardian*. https://www.theguardian.com/environment/ethicallivingblog/2008/mar/07/keepyourkettleincheck (2008).
12. Ligthart, T. & Ansems, A. Single Use Cups or Reusable (Coffee) Drinking Systems: An Environmental Comparison. *TNO Report R0246* (Netherlands Organization for Applied and Scientific Research, 2007).
13. Wallop, H. Reheat cold cups of tea, Government waste watchdog says. *The Telegraph*. https://www.telegraph.co.uk/news/earth/earthnews/6531540/Reheat-cold-cups-of-tea-Government-waste-watchdog-says.html (2009).

14. EAC. *Disposable Packaging: Coffee Cups*. House of Commons Environmental Audit Committee. https://publications.parliament.uk/pa/cm201719/cmselect/cmenvaud/657/657.pdf (2017).

15. Davis, A. & Moat, J. *Coffee Farming and Climate Change in Ethiopia: Impacts, Forecasts, Resilience and Opportunities*. Royal Botanic Gardens Kew and Environment & Coffee Forest Forum. https://www.kew.org/sites/default/files/Coffee Farming and Climate Change in Ethiopia.pdf (2017).

16. Gromko, D. Ethiopia's farmers fight devastating drought with land restoration. *The Guardian*. https://www.theguardian.com/sustainable-business/2016/may/02/ethiopia-famine-drought-land-restoration (2016).

17. Carbonbrief. Mapped: How every part of the world has warmed—And could continue to warm. *Carbonbrief.org*. https://www.carbonbrief.org/mapped-how-every-part-of-the-world-has-warmed-and-could-continue-to-warm (2018).

18. Bebber, D. P., Castillo, Á. D. & Gurr, S. J. Modelling coffee leaf rust risk in Colombia with climate reanalysis data. *Philos. Trans. R. Soc. B* **371**, 20150458 (2016).

19. Alemu, A. & Dufera, E. Climate smart coffee (*Coffea arabica*) production. *Am. J. Data Mini Knowl. Discov.* **2**, 62–68 (2017).

20. Jaramillo, J. *et al.* Some like it hot: The influence and implications of climate change on coffee berry borer (*Hypothenemus hampei*) and coffee production in East Africa. *PLoS One* **6**, e24528 (2011).

21. Stylianou, N. Coffee under threat. *BBC Online*. https://www.bbc.co.uk/news/resources/idt-fa38cb91-bdc0-4229-8cae-1d5c3b447337 (2017).

22. Gebreselassie, E. Caffeine high? Climate-hit Ethiopia shifts coffee uphill. *Reuters*. https://www.reuters.com/article/us-ethiopia-coffee-climatechange/caffeine-high-climate-hit-ethiopia-shifts-coffee-uphill-idUSKCN1J00ID (2018).

23. Moat, J. *et al.* Resilience potential of the Ethiopian coffee sector under climate change. *Nat. Plants* **3**, 17081 (2017).

24. Salamanca-Jimenez, A., Doane, T. A. & Horwath, W. R. Nitrogen use efficiency of coffee at the vegetative stage as influenced by fertilizer application method. *Front. Plant Sci.* **8**, 223 (2017).

25. Jayaraj, R., Megha, P. & Sreedev, P. Organochlorine pesticides, their toxic effects on living organisms and their fate in the environment. *Interdiscip. Toxicol.* **9**, 90–100 (2016).

26. FAO. *Phasing Out Highly Hazardous Pesticides is Possible! Farmer Experiences in Growing Coffee Without Endosulfan*. Food and Agriculture Organization of the United Nations. http://www.fao.org/3/a-i4573e.pdf (2015).

27. Bertrand, B. Breeding for the future: Coffee plants for the 21st century. *worldcoffeeresearch.org*. https://worldcoffeeresearch.org/work/breeding-future/ (2018).

28. Davis, A. P. *et al.* High extinction risk for wild coffee species and implications for coffee sector sustainability. *Sci. Adv.* **5**, eaav3473 (2019).
29. Ghosh, P. Satellites warn African farmers of pest infestations. *BBC Online.* https://www.bbc.co.uk/news/science-environment-46370601 (2018).
30. Haggar, J. & Schepp, K. *Coffee and Climate Change. Impacts and Options for Adaption in Brazil, Guatemala, Tanzania and Vietnam* (Climate Change, Agriculture and Natural Resource, 2012).

PART III

Lunch

CHAPTER 9

Climate-Smart Chicken

Abstract In the 1960s average yearly consumption of poultry meat stood at just over 3 kilograms per person on the planet. By the 1990s this had more than tripled, and by 2030 we are each predicted to be eating the equivalent of 17 kilograms of poultry meat a year (over 120 million tonnes worldwide). For every kilogram of chicken produced, up to 5 kilograms of greenhouse gas is emitted. Within this, producing the chicken feed is the main culprit at about three-quarters of the total. Each year in UK households, we waste 110,000 tonnes of chicken meat. In the developed world most chicken is produced under controlled conditions, so heat stress risks under a changing climate should be minimal. However, cooling capacities, transport systems and housing densities all need to be adapted as the risk of more extreme heat events increases. In the developing world chickens may be more exposed to severe weather impacts, but they also represent a useful way to enhance incomes and food security where resilient and higher-yielding varieties are made available.

Keywords Heat stress • Salmonella • Ventilation • Air-conditioning • Broilers • Shade • Transport • Free-range

Lunch is last night's chicken curry, scooped into a Tupperware container and re-heated in the office microwave. The bulk of the chicken we consume is produced in the same nation it is raised in—the availability of climate-controlled indoor housing methods meaning chicken can be produced on an industrial scale from the Arctic Circle to the equator.

Today's 50 billion-strong global chicken population is descended from just four species of jungle fowl domesticated in Southeast Asia over 4,000 years ago [1]. Most share a common ancestor in the so-called red junglefowl, but millennia of selective breeding have given rise to myriad varieties, from super egg layers like the Rhode Island Red, through sturdy meat chickens like the Cornish-Rock, to the poultry punk rockers that are Silkies—fluffy bundles that look like a cross between a chicken and a bag of pom-poms.

Chickens destined for the table are called 'broilers', and our appetite for their meat seems insatiable. In the 1960s average yearly consumption of poultry meat stood at just over 3 kilograms per person on the planet. By the 1990s this had more than tripled, and by 2030, we are each predicted to be eating the equivalent of 17 kilograms of poultry meat a year (over 120 million tonnes worldwide). The biggest producers of chicken—the US, Brazil and China—are also its most avid consumers (Fig. 9.1). Average

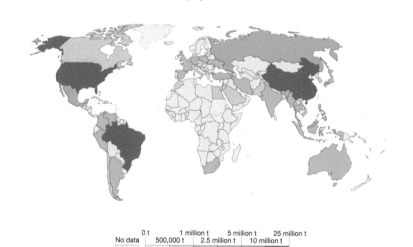

Chicken meat production, 2013
Annual production of chicken meat, measured in tonnes per year.

Our World in Data

| No data | 0 t | 1 million t | 5 million t | 25 million t |
| | 500,000 t | 2.5 million t | 10 million t |

Source: UN Food and Agriculture Organization (FAO)

CC BY

Fig. 9.1 Global chicken meat production in 2013 by country of origin (Source: Hannah Ritchie, Our World in Data) [3]. Available at: https://ourworldindata. org/grapher/chicken-meat-production

poultry meat consumption in the US is already a bucket-busting 40 kilograms per person a year [2].

The chicken meat in our own re-heated curry started life as a free-range bird raised in large flocks in England. Although free-range eggs have become popular over the last few decades (they now make up over half of those in UK stores), free-range chicken meat is still a niche commodity at under 5 per cent of UK production [4]. The bulk of chicken meat is instead produced in 'barn-raised' systems. These have a higher density—around 17 birds per square metre of sawdust-covered floor space—and, unlike free-range chickens, have no access to the outdoors. In both cases, commercial farmers aim to rush their poultry through—from new-born chick to oven-ready adult—in the most efficient way possible. Through careful selection of fast-growing traits combined with precisely controlled feeding regimes, modern farms are now able to grow a chicken to slaughter-ready size in under 40 days.

The chicks are hatched in incubators and for the first few weeks of life need warm and humid conditions, with the required temperature of around 32 degrees Celsius being achieved in controlled-atmosphere sheds. At one day of age, they are moved to the growing sheds and each is passed through a vaccination spray. The standard diet for barn-raised chickens consists of copious supplies of high-protein mash, crumbs or pellets. These are derived from a mix of cereal grains, like maize and wheat, along with vegetable and animal proteins like soy and fishmeal [5]. Various fats, minerals and vitamins are incorporated to boost energy contents and ensure good bone and muscle development. For free-range birds, a more varied diet including weeds and insects is possible. Controversially, many commercial chicken farmers also add antibiotics to feeds, along with pH control agents and enzymes to aid feed digestion [6].

Air-conditioned housing means temperature and humidity can be controlled to maximise growth and, for most commercial farms, this means keeping the chickens in windowless fan-ventilated barns. Good ventilation and thorough cleaning between batches are vital to avoid a dangerous build-up of ammonia from the poultry faeces—too much ammonia can lead to respiratory and eye damage [7]. Even under the controlled conditions of the broiler houses, an average of 3 in every 100 birds die during the rearing phase.

With a relatively short distance between the barn and our supermarkets, it is the on-farm production of broiler chickens that dominates their overall carbon footprint [8, 9]. For every kilogram of chicken produced, up to 5 kilograms of greenhouse gas is emitted. Within this, producing the chicken feed is the main culprit at about three-quarters of the total. During its short life, the average barn-raised table bird will consume more than double its own slaughter weight in feed. Free-range and organic birds consume closer to five times their final weight (as they have longer lives) and so, per drumstick, thigh and wing have a higher carbon footprint than their barn-raised cousins [10]. Soy meal from Brazil and Argentina is the main component of this feed and makes up almost half of its emissions, the rest coming from wheat, vegetable oils and fishmeal.

Given the heavy use of heating, ventilation and air conditioning by commercial producers, the use of electricity and gas are the next biggest source of emissions on the chicken farm. Housing, land use and manure complete the carbon footprint. It is, however, chicken manure's pungent plumes of ammonia that spread downwind of the farms that may pose the greatest environmental risk. Carried on the wind, the ammonia is deposited to fields, woodlands and lakes where it can damage sensitive species like lichens and mosses, and push up overall ecosystem nutrient levels (eutrophication). As an added bonus, it may boost nitrous oxide emissions too [11].

Once the chickens have reached market weight, they are transported to slaughterhouses, then to meat-processing plants, and on to retailers. Each of these phases adds a further 500 grams or so of emissions per kilogram of chicken produced, mainly through the fuel and energy used [8]. The final destination of the chickens, our households, appears as a relatively minor part of the total chicken meat climate impact. Refrigeration and cooking on average add just 350 grams of emissions per kilogram to the overall footprint. But, as with almost every type of food we consume, wastage then rears its ugly head.

For my siblings and me in 1970s' Britain, roast chicken was a special Sunday dinner treat. Ever hungry, our eyes grew wide as we watched our dad carefully carve the bird and lay slices of meat on our plates. The top privilege was to be given the Parson's nose—the fatty arrow-shaped base of the chicken's tail. Next most joyous (and often a recipe for stomach ache) was to clear your plate first and have prime picking of the carcass. Our small fingers became expert at peeling any remaining slivers of meat from the bones and nibbling the skin from wingtips.

Several decades of improved feeds, faster growth rates and industrial production methods have made chicken more of a staple than a Sunday treat. At the time of writing, a whole chicken can be bought in the UK for just £3. This incredibly low price has helped drive the huge increase in our consumption and along with it a lot more wastage.

Each year in UK households, we waste 110,000 tonnes of chicken meat. This is all avoidable. It is edible meat that doesn't get eaten because it has either gone past its use-by date (32,000 tonnes), because too much was cooked and served (46,000 tonnes), or simply because we didn't fancy chicken after all (13,000 tonnes) [12]. Some also ends up in the garbage because it is burned or spoiled. At 5 kilograms of greenhouse gas per kilogram of chicken, this Great British poultry wastage has a staggering carbon cost: half a million tonnes of emissions a year. It also means the equivalent of 85 million chickens are hatched, intensively reared, slaughtered, processed and sent to the shops, all to then end up in the bin [10].

A focus for cutting this poultry waste mountain down to size has been the huge amount that goes out of date before it can be eaten. Packaging can really help—separate compartments for chicken pieces, for instance, could slice 10,000 tonnes off British chicken waste simply by allowing us to use one side and freeze the other [13]. For whole birds, many chefs and recipe books have also been trying to help households make use of every last bit [14].

* * *

As chickens are produced all over the world, they face a wide array of climate risks. Heat is the big one for most, and the threats posed by higher temperatures can go far beyond the farm gate. From accelerated growth of dangerous spoilage bacteria like *Campylobacter* and *Salmonella*, to more barbecues and the inherent risks of an undercooked drumstick, hotter summers could ramp up the risks to our own health [15]. Climate-smart solutions must therefore make farms more resilient to severe weather impacts, lower carbon footprints, and better protect the health of birds and humans alike [16].

Food poisoning is unpleasant at the best of times. Poisoning from chicken meat can be deadly. The US Centers for Disease Control and Prevention estimate that around 1 million people become ill each year due to the eating of contaminated chicken [17]. With the surge in consumption, chicken has consequently become the meal most commonly involved in outbreaks of food poisoning. Poultry were associated with one quarter of the more than 1,000 outbreaks recorded in the US between 1998 and 2012. The bacteria *Salmonella* accounted for the most illnesses and hospitalisations, and the second highest number of deaths (*Listeria* being the number one food poisoning killer) [18]. Some of the more recent US outbreaks that involved high rates of hospitalisation have resulted from antibiotic-resistant strains of *Salmonella*.

Although contamination can occur during production and processing, many of the cases of food poisoning due to eating chicken are a result of poor storage, handling and undercooking [18]. Climate change, and especially higher temperatures, may accelerate the growth of spoilage organisms like *Salmonella*. Infections are already more common in summer and appear to peak during periods of warm weather. Where refrigeration is inadequate or the time between preparation and cooking is too long, the bacteria can multiply to dangerous levels. *Salmonella* poisoning commonly results in fever, diarrhoea and abdominal cramps [16]. In the UK there are an estimated 39,000 cases each year, but this represents a big decrease compared to the early 1990s. Right across Europe *Salmonella* poisoning appears to be on the decline thanks to improved food hygiene, better biosecurity, and the widespread vaccination of animals. Climate change is not expected to reverse this downward trend, though a 10 to 15 per cent increase in cases across Europe is possible by the 2080s under a very high warming scenario. In Australia, spiritual home of the barbecue, a more serious climate impact is likely. Here, an extra 4,000–7,000 cases of food poisoning per year by the middle of the century are predicted, and in South Australia, incidences of *Salmonella* may rise by more than 50 per cent [16].

With air-conditioned barns and carefully controlled atmospheres, most commercial chicken farms in the developed world might be expected to be at low risk from high temperatures and other direct climate change impacts [19]. Certainly, these systems are designed to protect the birds from weather extremes, giving as close to optimal conditions as possible to ensure maximum growth. Heat waves in the summer mean a rise in energy costs for cooling and ventilation, but warmer winters may cut winter heating bills. As climate change intensifies, it is the severity of the heat waves, and the risks of heat stress both in the barns and during transport, that represent one of the biggest threats—the 2003 European heat wave killed an estimated four million broiler chickens in France alone [20].

Today's commercial meat chicken is genetically distinct from those bred in the 1990s and before. It consumes a lot of feed, grows much faster and has a smaller heart relative to its body weight. It is therefore more sensitive to high temperatures and more susceptible to heart failure [21].

At the high stocking densities of commercial barns, the birds produce a lot of heat, and this needs to be actively managed. If the excess warmth is not efficiently removed, then the first signs of heat stress will quickly appear.

Initially a heat-stressed chicken will increase its water consumption, direct more blood supply to its comb, and try to move away from other birds. It may seek out cooler surfaces and areas where there is greater air-flow. Given room it will open its wings and try to expose more of its feather-free skin to the air. If it is still too hot the chicken will begin to pant and so shed more heat via evaporation. As the temperature rises still higher, so the rate of panting quickens and the bird becomes lethargic. With no respite it will die. Chickens like it warm—they have an internal temperature of 41 degrees Celsius—but an increase of 4 degrees Celsius or more above this is fatal [22].

As well as temperature, the humidity of the atmosphere in the broiler houses is a big risk factor. As humidity rises, the ability of the birds to cool themselves through evaporation falls. Damp and soiled bedding can exacerbate this humidity problem and lead to dangerous increases in ammonia in the barn air [23]. Even mild heat stress over an extended period is a serious economic issue for the farmer, as well as being a major animal welfare one. Heat-stressed chickens will eat less, have lower feed conversion efficiency, and so grow more slowly [24]. Their immune systems will also be suppressed, making them more susceptible to infection and disease.

The perfect storm of broiler heat stress impacts comes when high stocking densities combine with very hot weather and inadequate air-conditioning. Closely confined chickens are up to 40 per cent less efficient at shedding heat, and if the capacity of the barn's cooling and ventilation systems is exceeded, dangerous heating will occur. Farmers tend to plan their stocking densities months in advance, and so an unexpectedly intense summer heat wave may overwhelm the standard cooling systems they have in place [22].

The last leg of a meat chicken's life—that from barn to slaughter-house—adds further heat exposure risks. An estimated 1.8 million (2 in every 1,000) birds die in transit in the UK each year, with occasionally very high losses of more than 15 per cent reported. Heat stress plays a major role, with a pronounced peak in mortality during the summer months. Most birds are transported to slaughter in closely packed containers consisting either of loose crates or of modular stacks of metal drawers. Limited ventilation and space to move means the birds have less ability to keep themselves cool. As daily maximum outside temperatures rise above 17 degrees Celsius, so mortality rates tend to rise. Even in the low 20s Celsius, the mortality rate may double and above that it can increase more than sixfold [25].

Thankfully, large-scale chicken death incidents due to heat waves have tended to become less frequent in the developed world, with the use of improved barn designs, modern ventilation systems and improved transport practices. Tragedies still occur though, such as the loss of 50,000 chickens in North Carolina when power was lost for just one hour during the heat wave of 2011 [26] and the more than 6,000 birds that perished on a single farm in southern England in 2012 due to overstocking and inadequate ventilation during hot weather [27].

Often less apparent than these direct effects of extreme heat have been the effects of severe weather events on poultry feed. With feed being so reliant on staple crop ingredients like soy and wheat, severe weather events like droughts and heat waves can lead to shortages in supply and so sharp spikes in prices [28]. For chicken farmers in the developed world then, future climate change, and especially higher temperatures, poses risks to the welfare of their flocks and to productivity, as well as to feed and energy costs.

In the developing world the risks may be much greater. Here, access to modern housing and air-conditioning systems is often more limited. In some rural areas power supplies may be unreliable or non-existent. For commercial poultry farmers in already-hot regions of Africa and Asia, concern is growing that climate change and more intense heat waves will increasingly threaten livelihoods and the welfare of birds and humans alike [29].

Many smallholders in rural areas use chickens as an extra source of protein and income to supplement their normal farming. Their birds are more likely to be free ranging and the housing to have no active cooling and ventilation. Most chickens will scavenge for food rather than being given feed, and in general, their growth rates are low compared to birds on commercial farms. They are also more prone to disease, parasites, and to being attacked by predators [19]. During periods of intense heat, free-range chickens will usually respond by seeking shade. However, if this shade-retreat is for prolonged periods and no additional feed is supplied, then their nutrition may suffer. In some areas changing precipitation and higher temperatures may also combine to reduce the availability of food that can be scavenged by the chickens as they roam. With food and income from chickens often providing the final safety net from poverty for rural smallholders, avoiding reduced productivity and increased mortality due to the impacts of climate change can yield big social benefits.

*   *   *

Achieving climate-smart chicken production inevitably requires context-specific approaches. For large commercial units in the UK and other developed nations, planning ahead for extremes of heat and ensuring that housing, transport and management practices are fit for purpose are at the core. While close attention to forecasts and severe weather warnings can help give a few days to check cooling and ventilation before a heat wave hits, longer term planning for lower stocking densities in the summer months may be much more effective. Output from the farms may be lower, but they can better avoid the heat stress risks caused by too many chickens and very high temperatures coinciding. Some farmers already thin out their numbers in expectation of hot weather to allow the remaining birds to keep cool. Many have improved the insulation and thermal performance of their barns, taking account of its positioning in the landscape and how reflective the roof materials are [30]. Good insulation has the multiple benefits of lowering heating costs in winter, buffering extreme outside heat in summer, and reducing condensation and damp-related issues inside the barns.

Making use of natural shade from trees may also be beneficial during the hottest days—this is especially true for free-range systems where the birds need areas of shade they can retreat to in the heat of the day. So too can using narrow barns (to aid ventilation air flow) and ensuring that air intakes for ventilation and cooling are located in the best locations (such as shaded walls) to draw in cool air.

The features of the landscape surrounding the chicken farm can play a much wider role in just how climate-smart it is. Poorly maintained access roads, inadequate manure storage and leaky grain silos all make the farm more vulnerable to storms and intense rainfall events—flooding having the potential to block roads, spoil feed, and cause pollution of streams and rivers through surface runoff. To limit the plumes of ammonia gas emitted from commercial broiler farms, good manure management is key. This involves regular cleaning of the sheds and treatment of soiled bedding to make it more acidic (so reducing how much ammonia is lost to the air). Strategic planting of tree 'capture belts' downwind of the broiler sheds can also be an effective way to intercept the ammonia carried on the wind. In some cases this type of 'poultry agroforestry' has the potential to boost tree growth, and so carbon uptake, of the farm too [31].

Building design and location features can therefore help reduce the overall amount of mechanical cooling and ventilation required, but in commercial systems some kind of air-conditioning system will still be needed. During hot weather an effective ventilation system keeps the air over the chickens moving, taking with it heat and moisture that could otherwise build up and cause heat stress problems. It means that, when the birds pant, their evaporative cooling is much more effective with the feeling of wind chill from the air movement also serving to make the birds more comfortable.

Many older barns do not have such active cooling systems and rely more on natural ventilation via openings along the length of the barn. In very hot weather internal fans that circulate the air more can help make the birds feel more comfortable, though these fans do not provide new air from outside and so risk a dangerous build-up of moisture, carbon dioxide and ammonia in the barn's atmosphere [22].

To make the most of the benefits of evaporative cooling, misting sprays are sometimes used. These can be a very effective way of avoiding the panting response of the birds to high temperatures, and so keeping feeding and growth rates going strong. The downside of these water spray systems is that, at very high temperatures, they can actually make things worse—raising the humidity to a level where the birds cannot cool themselves at all. Most commercial farms now use alarm systems that warn when cooling and ventilation equipment fail, or when temperature and humidity are approaching dangerous levels. The alarms either trigger automatic back-up cooling or warn the farmer to start up failsafe plans (like the use of generators during a power cut) [22].

Day to day, good practice during hot weather periods depends very much on the expertise of the commercial chicken farmer. For instance, leaving the ventilation system running all night may seem a waste of money and energy, but then these costs may pay big dividends in terms of greater comfort and survival rates of the birds during the heat of the following day. Likewise, reducing or completely removing feed during the hottest hours of the day could reduce growth rates, but may also limit overheating in the birds and the numbers which then succumb to heat stress [22].

The capture and onward transport of broiler chickens to the slaughterhouse can be a very stressful experience and, as we've seen, there remains a real danger of heat stress during this final stage of the chicken's life. Rapid capture and transport coinciding with the coolest part of the day helps reduce heat stress risks, and postponing things until the weather cools and the birds are in a less stressed state can also pay big dividends. During transport, ensuring proper ventilation, drinking water and adequate space are a must, as is avoiding the travel modules and their vehicle being left standing in direct sunlight.

All of these responses have the potential to yield greater resilience to extreme weather, can reduce heat stress risks and mortality rates, and so boost productivity. Through better hen welfare, the energy and feed inputs—representing the largest part of a broiler chicken's carbon footprint—will also be more efficient. Implementing some of the technical solutions, like mechanical ventilation and super-insulated housing, requires significant upfront costs, and so, the availability of finance is important. More crucially, informed planning and practices to deal with increasing heat stress risks require good training and ready access to advice on weather forecasts and best practice.

For millions of rural smallholders around the world, chickens represent an opportunity to boost incomes and the overall resilience of households. In Nyando—a largely rural district of Kenya—keeping chickens has actually become a cornerstone of climate-smart activities for many families [32]. The average household keeps 30 birds that largely scavenge for their food and are indigenous to the area (and so can thrive even with minimal feed and healthcare). An increasing number of women in these households are now farming a crossbred type of chicken that matures faster than the indigenous ones (they reach market weight in one-third of the time). Combined with training on the construction and maintenance of better housing, provision of feed, and more disease control and treatment, the scheme is already proving a success. This innovative 'climate-smart chicken' initiative is enabling the women of Nyando to build more resilience into their livelihoods even as the threats to their other crops from climate change increase—the chickens represent a feathered form of household savings and insurance against times of heavy crop losses, with the manure they produce being a valuable fertiliser to boot.

So, the climate future of the world's chicken dinners appears safe, albeit one where the welfare of the birds and safety of their meat requires renewed attention. A bigger question is to what extent the staggering growth in chicken meat consumption in the twentieth century will continue through the twenty-first. The push for more sustainable diets is certainly gaining momentum, with very high-carbon meats like beef and lamb sitting right in the emissions reduction bullseye. Vegetarianism and veganism may be gaining ground, but there's little sign that we'll be turning our backs on the finger-licking delights of chicken any time soon.

REFERENCES

1. PennState. *History of the Chicken.* Pennsylvania State University Extension. https://extension.psu.edu/history-of-the-chicken (2011).
2. thepoultrysite.com. *Global Poultry Trends 2014: Growth in Chicken Consumption in Americas Slows.* http://www.thepoultrysite.com/articles/3324/global-poultry-trends-2014-growth-in-chicken-consumption-in-americas-slows/ (2015).
3. Ritchie, H. Global chicken meat production, 2013. *Ourworldindata.org.* https://ourworldindata.org/grapher/chicken-meat-production (2018).
4. Castella, T. Do people know where their chicken comes from? *BBC Online.* https://www.bbc.co.uk/news/magazine-29219843 (2014).
5. poultryhub.com. *Feed Ingredients.* http://www.poultryhub.org/nutrition/feed-ingredients/ (2019).
6. poultryhub.com. *Feed Additives.* http://www.poultryhub.org/nutrition/feed-ingredients/feed-additives/ (2019).
7. Aziz, T. & Barnes, H. J. Harmful effects of ammonia on birds. *Poultry World.* https://www.poultryworld.net/Breeders/Health/2010/10/Harmful-effects-of-ammonia-on-birds-WP008071W/ (2010).
8. Skunca, D., Tomasevic, I., Nastasijevic, I., Tomovic, V. & Djekic, I. Life cycle assessment of the chicken meat chain. *J. Clean. Prod.* **184**, 440–450 (2018).
9. MacLeod, M. *et al. Greenhouse Gas Emissions from Pig and Chicken Supply Chains—A Global Life Cycle Assessment*, 171 (Food and Agriculture Organization of the United Nations (FAO), Rome, 2013).
10. Leinonen, I., Williams, A., Wiseman, J., Guy, J. & Kyriazakis, I. Predicting the environmental impacts of chicken systems in the United Kingdom through a life cycle assessment: Broiler production systems. *Poult. Sci.* **91**, 8–25 (2012).
11. Reay, D. *Nitrogen and Climate Change: An Explosive Story* (Springer, 2015).
12. WRAP. Household food and drink waste in the United Kingdom 2012. *Waste and Resource Action Programme.* http://www.wrap.org.uk/sites/files/wrap/hhfdw-2012-main.pdf.pdf (2013).
13. WRAP. Packaging design to reduce household meat waste. *Waste & Resources Action Programme.* http://www.wrap.org.uk/sites/files/wrap/packaging_design_to_reduce.pdf (2012).
14. Seal, R. How to make a chicken last a week. *The Observer.* https://www.theguardian.com/lifeandstyle/2017/apr/09/how-to-make-a-chicken-last-a-week-angela-hartnett (2017).

15. Lake, I. *et al.* *Food and Climate Change: A Review of the Effects of Climate Change on Food Within the Remit of the Food Standards Agency* (Food Standards Agency, 2010).

16. Lake, I. R. Food-borne disease and climate change in the United Kingdom. *Environ. Health* **16**, 117 (2017).

17. CDC. *Chicken and Food Poisoning*. https://www.cdc.gov/features/salmonellachicken/index.html (2018).

18. Chai, S., Cole, D., Nisler, A. & Mahon, B. E. Poultry: The most common food in outbreaks with known pathogens, United States, 1998–2012. *Epidemiol. Infect.* **145**, 316–325 (2017).

19. Nyoni, N., Grab, S. & Archer, E. R. Heat stress and chickens: Climate risk effects on rural poultry farming in low-income countries. *Clim. Dev.*, 1–8 (2018).

20. García-Herrera, R., Díaz, J., Trigo, R. M., Luterbacher, J. & Fischer, E. M. A review of the European summer heat wave of 2003. *Crit. Rev. Environ. Sci. Technol.* **40**, 267–306 (2010).

21. Zhang, J., Schmidt, C. J. & Lamont, S. J. Gene expression response to heat stress in two broiler lines. *Anim. Ind. Rep.* **662**, 61 (2016).

22. DEFRA. *Heat Stress in Poultry: Solving the Problem*. Department for Environment, Food & Rural Affairs, UK. https://assets.publishing.service.gov.uk/government/uploads/system/uploads/attachment_data/file/69373/pb10543-heat-stress-050330.pdf (2005).

23. PoultryWorld. *Harmful Effects of Ammonia on Birds*. https://www.poultryworld.net/Breeders/Health/2010/10/Harmful-effects-of-ammonia-on-birds-WP008071W/ (2010).

24. Lara, L. & Rostagno, M. Impact of heat stress on poultry production. *Animals* **3**, 356-369 (2013).

25. Warriss, P., Pagazaurtundua, A. & Brown, S. Relationship between maximum daily temperature and mortality of broiler chickens during transport and lairage. *Br. Poult. Sci.* **46**, 647–651 (2005).

26. Hegeman, R. Heat wave kills thousands of poultry. *Huffington Post*. https://www.huffingtonpost.com/2011/07/13/heat-wave-poultry_n_896812.html (2017).

27. BBC. *Paul Flatman Banned after Chelmsford Chicken Heat Deaths*. https://www.bbc.co.uk/news/uk-england-essex-32847446 (2015).

28. Hill, C. Free range egg farmers battle rising feed costs after hot, dry summer. *Eastern Daily Press*. http://www.edp24.co.uk/business/farming/free-range-poultry-rising-feed-costs-bfrepa-1-5655114 (2018).

29. Bhadauria, P. *et al.* Impact of hot climate on poultry production system-a review. *J. Poult. Sci. Tech.* **2**, 56–63 (2014).

30. Fairchild, B., Vest, L. & Tyson, B. L. Key factors for poultry house ventilation. *thepoultrysite.com.* http://www.thepoultrysite.com/articles/2321/key-factors-for-poultry-house-ventilation/ (2012).

31. Bealey, W. *et al.* The potential for tree planting strategies to reduce local and regional ecosystem impacts of agricultural ammonia emissions. *J. Environ. Manag.* **165**, 106–116 (2016).
32. Recha, J. & Kimeli, P. *Chicken to the Rescue: How Farmers in Nyando are Managing Climate Risks.* CGIAR Climate Change, Agriculture & Food Security. https://ccafs.cgiar.org/blog/chicken-rescue-how-farmers-nyando-are-managing-climate-risks#.XOxQjS2ZPUo (2017).

CHAPTER 10

Climate-Smart Rice

Abstract Rice is grown in a huge range of locations across over 160 million hectares of the planet, from the cool temperate regions of Northeast Asia, through low-lying river deltas in the tropics, to lofty Himalayan mountain slopes at altitudes of over 2 kilometres. To produce each kilogram bag of Basmati rice, the climate-warming equivalent of over a kilogram of carbon dioxide (mainly in the form of methane) is also emitted. Along with some nitrous oxide from the use of nitrogen fertilisers on the fields, the life-cycle emissions can top 1.5 kilograms per kilogram. In Britain rice wastage amounts to over 40,000 tonnes each year and so the equivalent of around 60,000 tonnes of greenhouse gas emissions. Drought, flood, heat and disease are all major risks that may be exacerbated by climate change. Loss of irrigation water supplies due to melting of glaciers in the Himalayas is a major concern for rice growers in South Asia. Improved water management can give greater resilience to climate change and radically reduce methane emissions at the same time.

Keywords Basmati • Punjab • Pakistan • India • Methane • Methanogens • Alternate wetting and drying • Drainage • Rice straw • Puddling • Irrigation • Himalayas • Glacial melt

Along with last night's chicken curry is a portion of leftover Basmati rice. Each year we consume over 500 million tonnes of this unassuming grain. It is the main source of nourishment for over a billion people and has been at the heart of global food security for centuries—rice has been found at archaeological sites dating as far back as 8,000 BC. Today there are two main domesticated types: *Oryza sativa* (known as Asian rice) and *Oryza glaberima* (or African rice) [1]. Within these groups, however, there are more than 4,000 varieties—even Basmati has its own cadre of forms, from Super Basmati (long grained and aromatic) to Shaheen Basmati (a salt-tolerant variety grown in areas where brine has contaminated ground waters) [2].

Rice is grown in a huge range of locations across over 160 million hectares of the planet, from the cool temperate regions of Northeast Asia, through low-lying river deltas in the tropics, to lofty Himalayan mountain slopes at altitudes of over 2 kilometres. What all these growing areas have in common is a good supply of water [3].

The goliaths of rice production, and consumption, are China and India. Together they make up more than half of the global harvest, amounting to over 350 million tonnes of rice each year [4] (Fig. 10.1). India and

Rice production, 2014
Annual paddy rice production, measured in tonnes per year.

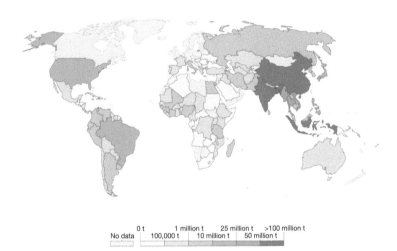

| No data | 0 t | 1 million t | 25 million t | >100 million t |
| | 100,000 t | 10 million t | 50 million t | |

Source: UN Food and Agriculture Organization (FAO)   CC BY

Fig. 10.1 Global rice production in 2014 by country of origin (Source: Hannah Ritchie, Our World in Data) [6]. Available at: https://ourworldindata.org/grapher/rice-production

Thailand lead the global pack in terms of exports to big foreign markets like the European Union and the US. For our own re-warmed Basmati, the heartlands of production are found in northwestern India and eastern Pakistan, especially the humid, sub-tropical expanses of Punjab [5].

Basmati rice grown in the fields of Indian and Pakistani Punjab is well known for long, slender grains and its popcorn-like fragrance—the name Basmati translates from Hindi as queen of fragrance [2]. It owes its beguiling aroma to high levels of aromatic compounds, and Basmati commands a premium market price as a result [7]. The plants need prolonged sunshine, high humidity and plenty of water, so thrive in Punjab's warm and wet sub-tropical climate, where high temperatures early in the year give way to a warm and humid rainy season from June to September.

Most rice is first grown as seedlings in nursery plots for about a month and then transplanted to the waiting fields. The flooded paddy soils are prepared by puddling—dragging a harrow up and down through the water-covered soils to stir up clay particles that then block pores in the soil. Once the seedlings are planted, the puddling means more water is retained, weeds are killed off and the young rice plants can begin growing with their roots fully immersed.

In the irrigated systems common to Punjab, the rice paddies are kept submerged for 2 to 4 weeks after planting and then irrigated again as the water drains into the soil. Nitrogen and phosphorous fertilisers are added (ideally at times when the fields are not flooded) to boost plant growth, along with fungicides and insecticides. If all goes well then the grain-heavy rice plants should be ready for harvest within about 4 months. Fields are allowed to drain prior to harvest in October and November so as to make collection easier and have the soil ready for the planting of a follow-on crop. The quality of rice is very sensitive to temperature—cooler, slower growth tends to mean better cooking and taste, while hotter temperatures can mean the rice produced becomes sticky during cooking and loses its much sought-after flavour [8].

After harvest the rice is threshed and cleaned to separate good grain from straw and immature seeds, before being bagged up for transport and processing. If the moisture content is high (above about 15 per cent), the rice is also dried to reduce spoilage risks. For the perfect white rice we know and love in Britain, the grains are milled. The rice husks, any soil particles and the dark outer layers of the grains—the bran—are removed (for brown rice, this bran layer is retained). Individual grains of white rice are then polished and graded by size to filter out any that are broken or misshapen [9]. Perfect whole grains of Basmati rice are the premium result of all this processing, but little is wasted—the husks are often used for fuel, the bran is used for animal feed and to make bran oil, and the broken rice fragments (called brewers' rice) are used for making beer, rice flour or noodles. If the harvest is destined for international markets like the UK, it travels by truck to the coast and is then shipped from large ports like Karachi.

Harvesting and processing, transport and cooking, each contribute to the total carbon footprint of the rice we consume. These emissions come largely from fossil fuel use and add up to around 80 grams of carbon dioxide for each kilogram of rice produced [10]. As with most foods, however, it's the growing of the rice in the first place that dominates its life-cycle emissions. Waterlogged fields like those in which Basmati is often grown are the perfect home for that climate-changing group of microbes that is the 'methanogens'. Just as these microbial methane–producers enjoy the warm, low-oxygen and carbon-rich environs of dairy cow rumens (Chap. 5), so they can thrive in the muddy sediments of rice paddies. Paddy soils overlain for long periods with standing water provide exactly the conditions they need to proliferate—the methane gas then either diffuses up through the overlying water and sediment, is released via bubbles, or finds its way up the stems of the plants and out into the air. With huge areas of land under cultivation, and soils that are often methane-producing hot spots, rice is a major source of human-induced methane emissions worldwide [11].

To produce each kilogram bag of Basmati rice, the climate-warming equivalent of over a kilogram of carbon dioxide (mainly in the form of methane) is also emitted. Along with some nitrous oxide from the use of nitrogen fertilisers on the fields, the life-cycle emissions can top 1.5 kilograms per kilogram of rice [10]. This footprint means that any wastage has a significant climate cost too. Spoilage due to the harvest being too moist or badly stored can be a problem, as can water damage and poor ventilation during the long journey by sea to foreign markets [12]. For consumers, the fact that the rice we buy is usually well dried and packaged means relatively little goes past its use-by date. Much more common is wastage from preparing and serving too much.

In Britain such avoidable waste amounts to over 40,000 tonnes of rice each year [13] and so the equivalent of around 60,000 tonnes of greenhouse gas emissions. Cooking only what is required, serving only what is wanted and using any leftover cooked rice for other meals are again the leading responses. In truth, judging just the right amount of rice can be tricky—its huge expansion when cooked often means an amount that looked about right in dried form turns into a daunting heap at the dinner table. Measuring cups can help, and if the inevitable overload still happens, then at least rice is great for combining with other meals. Once cooked though, care must be taken that leftovers are well refrigerated and consumed fairly quickly—bacterial growth in cooked rice can be rapid, and so food poisoning becomes a risk if it is left uneaten for too long [14].

*   *   *

With rice being grown in so many different areas around the world, the threats its farmers face from severe weather events, pests and diseases are manifold. The Basmati growers of the Punjab are all too aware of most of them. Since the 1960s maximum spring temperatures have increased markedly. Alongside this has come higher rainfall, sometimes with disastrous effects [15].

Rice plants like it wet, but extreme rainfall events can destroy the retaining banks around paddies, erode their soils and, if submerged too long, kill the plants too [16]. Across Asia, millions of tonnes of rice are lost each year due to flooding. In July and August of 2010 heavy and prolonged rainfall caused devastation in Pakistan, leaving one-fifth of the land area submerged [17]. These floods claimed the lives of over 1,600 people and forced two million from their homes. In the Punjab over half a million hectares of cropland were inundated. Wheat, cotton and sugar farming were all badly hit, and an estimated 200 thousand tonnes of rice were lost [18]. The wet season of 2014 again saw record-breaking rainfall across the Punjab, with Lahore enduring 300 millimetres of rain (around half of the annual average) in just 24 hours.

As the quality of Basmati rice is very dependent on temperature—the plants need a relatively cool period of growth to fill their 'ears' with the largest, best-tasting grains—heat waves can also spell disaster. The hottest period in the Punjab tends to be from May to June, when daytime temperatures above 40 degrees Celsius are the norm. Pakistan is no stranger to heat waves, experiencing an average of 7 every year [19] (in 2010 the southern city of Mohenjo Daro re-wrote the Asian record books with a temperature of 53.5 degrees Celsius). For Pakistani Punjab such extreme temperatures are rare, but the thermometer often pushes past

45 degrees Celsius in its capital, Lahore, and in May 2018, rice-growing epicentres on both sides of the Pakistan-India border were hit [20]. Climate change is set to enhance both the frequency and severity of such heat waves. Under a high-emissions scenario, the chance of heat waves in Punjab may increase by one-third by as early as 2030. By the end of the century, it could more than double [21].

The risks of climate change to Punjabi rice growers are already apparent. Warming is approaching critical levels in the Basmati-growing heartlands due to it coinciding with times when the rice plants are at their most susceptible [22]. By the second half of the century, annual average temperatures in the region could increase by more than 3 degrees Celsius [23]. As each degree of night-time warming can cut rice yields by 10 per cent [24], such a rise could see rice harvests fall by more than a quarter.

For rainfall the outlook appears brighter, average amounts falling each year are not expected to change much in the coming decades. Instead, it is more variability in where and when rain occurs that risks causing damage through torrential downpours and flash flooding. Where this same variability means more drought, the fact that nine out of ten rice fields in the Punjab have irrigation should help give resilience. For many thousands of rice farmers in South Asia, however, it's the impact of warming on the distant frozen source of this vital irrigation water that is really focussing minds.

Both the Indus and Ganges rivers derive much of their dry season flow from Himalayan glaciers. The Indus—a river whose waters irrigate the crops on which more than half of the Pakistani population depends—will see lower flows as the glaciers high up in the Himalayas succumb to rising temperatures. By the middle of the century melt waters from these lofty peaks may have diminished by a third. The timing of the melt could make things even worse. As temperatures increase, so the glacial waters are released ever earlier in the year. For South Asia this means that river tributaries relied on for irrigation water may flow much faster during the wet season (when irrigation is less crucial) but then trickle to a halt in the dry season, just when they are needed most. Lower river flows also mean ground waters and aquifers are recharged more slowly—in north western India and eastern Pakistan, water tables are already falling and wells drying up [22].

Ironically, as water shortages become ever more likely for the rice growers of Asia, one saving grace could come in the form of more carbon dioxide. The same increases in carbon dioxide concentrations that are driving up global temperatures also make for faster plant growth and so, potentially, bigger rice yields. A doubling of concentrations in the atmosphere could boost harvests by a quarter—the extra carbon dioxide also means water demands of plants are reduced (helping buffer drought impacts). The climate change impacts that come with a 2-degrees-Celsius temperature rise, however, would cancel out much, if not all, of this positive carbon dioxide fertilisation effect [25].

Lurking as a diminutive threat that could take advantage of all these changes are the myriad pests and diseases that beset rice plants. Basmati is especially susceptible to stem borers. These pernicious insects feed on the still-forming tillers of rice (the stems that should go on to carry seed) and can devastate harvests—the striped stem borer common to Asia is able to wipe out whole crops [26], and in India stem borers have been putting holes in yields for decades [27]. Leaf hoppers also cause a lot of damage in rice paddies. Attack by high populations of these tiny insects causes the rice leaves first to turn yellow-orange, and then to die. As they feed, leaf hoppers may transmit incurable diseases like the beautifully descriptive Rice Ragged Stunt [28]. Higher temperatures appear to be a boon for rice pests like the Brown Leaf Hopper, as they reduce winter mortality rates and, in the hottest months, may kill off natural predators and so allow pest populations to soar [29]. A warmer winter season is also expected to benefit leaf blast, a fungal disease that attacks rice leaves causing a scattering of grey-green lesions that gradually expand across the entire leaf surface until it is dead [30]. Where higher carbon dioxide concentrations boost plant growth, diseases like leaf blast can take even greater advantage, as the lusher rice leaves appear to become more susceptible to invasion and destruction [31]. Rice farmers around the world already lose over a third of their crop to pests and diseases [32], so a climate boost to all the borers, hoppers and stunts is the last thing they need.

*   *   *

For the rice growers of Pakistan, including those that provided my own lunchtime Basmati, the many interactions of a changing climate by the middle of the century are predicted to increase the area on which rice is grown by around 2 per cent, but to simultaneously cut yields by almost 6 per cent [17]. At the heart of climate-smart responses to the increasing threats posed by flood, drought and faltering Himalayan melt water supplies is water management. Here, a system of alternate wetting and drying is already used to boost resilience. About two weeks after trans-planting of the seedlings, the fields are allowed to dry out. Farmers then closely monitor how saturated the paddy soils are using a simple pipe (called a pani) that has holes drilled along its bottom half. The pani-pipe is pushed vertically down and the soil inside is scooped out so that the saturation of the field can be easily checked by looking down the pipe to see what level the water table is sitting at [33]. When it drops too far the field is flooded again. This simple monitoring means farmers can forego constant flooding and instead inundate the rice only during the critical growth stages, such as flowering. By better matching the needs of the plants, and so saving water, they can reduce vulnerability to droughts and to faltering irrigation sources [17]. Along with greater climate change resilience comes a major bonus in the form of reduced methane. Periods of alternating wet and dry conditions mean that more oxygen can penetrate the soils and the multitudes of methanogens they contain are knocked back. Over the course of the rice-growing season, this prac-tice can halve methane emissions [34], and so slice as much as one-third off the total lifecycle carbon footprint of the rice produced.

Yields can benefit from such improved water management too. Growers from right across Asia report a boost to harvests and lower losses from lodging—where plant stems are bent over by wind or rain [35]. Lodging makes the rice harder to harvest and more prone to pest and disease attacks (rice plants have short roots and are especially vulnerable). Periods of flood-free soil mean the plants establish deeper roots, are better at resist-ing the pushes and pulls of strong winds and torrential downpours, and are more able to cope with any intense droughts that occur [36].

As a climate-smart approach, improved water management in rice agri-culture has become somewhat of a poster boy. There are some caveats, such as slight increases in emissions of nitrous oxide (allowing more oxy-gen into the soil promotes its production) and the potential for more weed problems (constant flooding helps control weeds), but time and again, the approach has delivered on the magic triumvirate of bigger har-vests, greater climate resilience and lower emissions.

Still more climate benefits can be gained by careful management of rice straw after harvest and the timing of manure, compost and fertiliser applications. For rice straw, dropping piles of freshly cut stalks into the stagnant water of the paddies provides a feast for the methanogens. Composting it at the side of the field instead, and then applying it back to dry soils during the off-season, means much less methane is produced and soil carbon is enhanced. Some farms even turn the methanogens into an ally, collecting together the rice straw and any available manure and allowing this carbon-rich mix to decompose in sealed digesters where the methane produced is then collected and used as fuel [37].

Like water, closely matching the use of fertilisers with the needs of the plants as they grow can bring big dividends. Too little nitrogen, for instance, and the plants will be stunted, turn yellow and produce little grain. Too much, and the excess nitrogen will pollute drainage streams and increase emissions of ammonia and nitrous oxide to the atmosphere. Leaf colour charts—simple sheets printed with four coloured strips in different shades of green—are commonly used to allow farmers to gauge whether their crop is looking too yellow and so needs a fertiliser top-up [38]. As every field and crop is different, site-specific nutrient management is much more effective than a one-size-fits-all approach. Alongside leaf colour charts, this can involve the calculation of the optimum water and fertiliser inputs, and their timings, based on the climate, soil type and management at each location. Trials by farmers in India have shown boosts in harvests of almost a tonne of rice and for each hectare where this approach is used [39].

Arguably the most powerful tool at our disposal to resist the climate-powered march of drought and flood, of pests and diseases, are the rice plants themselves. Rice has been the focus of intensive research and breeding programmes for decades, yielding famous varieties like 'Golden Rice'—a genetically modified plant intended to help address vitamin A deficiency in children and pregnant women [40]. Others have been developed to have shorter stems (to reduce lodging) and to be more drought, flood or even salt tolerant—rising sea levels are pushing saltwater further and further inland [16].

In areas where a particular pest or disease is rife, planting a rice variety that has good resistance to it is often the most cost-effective first line of defence. Good follow-up practices include regular cleaning of equipment, stores and fields to avoid disease spread, avoiding over-application of fertilisers (this makes the plants more susceptible to attack) and encouraging natural pest predators by providing habitats and limiting pesticide use [32]. Part of a climate-smart response is planning (and planting) ahead for new pest and disease risks as temperatures rise and rainfall patterns change. It also requires understanding of the ways in which severe weather events could undermine traditional controls—for example, more intense rainfall washing pesticides into drainage waters.

Of course, having a wealth of rice varieties to better fit with local conditions and future climates is of little use if they are not accessible to farmers or the climate change impact projections are too uncertain. Organisations like the International Rice Research Institute are actively translating the latest knowledge on plant breeding and rice farm management into practice around the world. At the local level, well-funded extension services are also needed to provide a gateway to finance and technology and allow farmers to learn about solutions like alternate wetting and drying, site-specific nutrient management and the rest.

Sharing of best practice and expert advice through farmer field schools has already led to some rice growers in Afghanistan doubling their yields while at the same time slashing their water use, fertiliser inputs and incidences of pest and disease attack [41]. In Pakistani Punjab, an increasing number of farmers are making use of laser levelling technology to ensure rice fields are flat, water leakages are reduced, and patches that will become too wet or dry are avoided. This relatively simple technology (a laser beam is fired from a box on the side of the field and hits a receiver on the farmer's plough to guide its depth) is used during preparation of a rice paddy, but so far, its wider uptake has been hobbled by a lack of funding [17].

The enormous scale of rice production and consumption means that achieving a climate-smart future for this mealtime staple has implications for food security worldwide. The Basmati growers of Punjab are in the vanguard of farmers trying to boost yields and the resilience of their crops while at the same time cutting emissions. The devastating floods of 2010 left some 90 million people in Pakistan without a secure food supply [8]; climate-smart approaches, including those for rice, could help avoid even bigger threats to food security in the coming decades.

REFERENCES

1. Sweeney, M. & McCouch, S. The complex history of the domestication of rice. *Ann. Bot.* **100**, 951–957 (2007).
2. Ashfaq, M. Basmati rice a class apart (A review). *Rice Res.: Open Access* (2014).
3. Bouman, B. How much water does rice use. *Management* **69**, 115–133 (2009).
4. EC. *EU Rice Economic Fact Sheet.* European Commission. https://ec.europa.eu/agriculture/sites/agriculture/files/cereals/trade/rice/economic-fact-sheet_en.pdf (2015).
5. APEDA. *Basmati Survey Report.* APEDA, New Delhi. http://apeda.gov.in/apedawebsite/six_head_product/BSK-2017/Basmati_Report-1.pdf (2017).

6. Ritchie, H. Global rice production, 2014. *Ourworldindata.org*. https://ourworldindata.org/grapher/rice-production (2018).

7. Gaur, A. *et al.* Understanding the fragrance in rice. *Rice Res.: Open Access* (2016).

8. AGRIPB. *Cultivation of Basmati Rice.* http://agripb.gov.in/pub/pdf/cultivation_of_basmati_rice.pdf (2018).

9. IRRI. *Rice Knowledge Bank—Milling.* http://www.knowledgebank.irri.org/step-by-step-production/postharvest/milling (2018).

10. Pathak, H., Jain, N., Bhatia, A., Patel, J. & Aggarwal, P. K. Carbon footprints of Indian food items. *Agri. Ecosyst. Environ.* **139**, 66–73 (2010).

11. Reay, D. S., Smith, P., Christensen, T. R., James, R. H. & Clark, H. Methane and global environmental change. *Annu. Rev. Environ. Resour.* **43**, 165–192 (2018).

12. Safety4Sea. *Transportation of Rice Cargo.* https://safety4sea.com/transportation-of-rice-cargo/ (2013).

13. WRAP. Household food and drink waste in the United Kingdom 2012. *Waste and Resource Action Programme.* http://www.wrap.org.uk/sites/files/wrap/hhfdw-2012-main.pdf.pdf (2013).

14. Kelly, R. 22 recipe ideas for leftover rice. *The Guardian.* https://www.theguardian.com/lifeandstyle/2014/jun/03/22-recipe-ideas-for-leftover-rice (2014).

15. Khattak, M. S. & Ali, S. Assessment of temperature and rainfall trends in Punjab province of Pakistan for the period 1961–2014. *J. Himal. Earth Sci.* **48**, 42–61 (2015).

16. IRRI. *Climate Change-Ready Rice.* http://www.knowledgebank.irri.org/step-by-step-production/pre-planting/rice-varieties/item/climate-change-ready-rice (2018).

17. CIAT. Climate-smart agriculture in Pakistan. *CSA Country Profiles for Asia Series.* International Center for Tropical Agriculture (CIAT); The World Bank. Washington, DC. 28 p. http://sdwebx.worldbank.org/climateportal/doc/agricultureProfiles/CSA-in-Pakistan.pdf (2017).

18. Anthony, A. & Georgy, M. Pakistan floods destroy crops and could cost billions. *Reuters.* https://www.reuters.com/article/us-pakistan-floods-agriculture/pakistan-floods-destroy-crops-and-could-cost-billions-idUSTRE67B0EW20100812 (2010).

19. Shaikh, S. Pakistan faces increasing heat waves. *Scidev.net.* https://www.scidev.net/asia-pacific/climate-change/news/pakistan-faces-increasing-heat-waves.html (2018).

20. Timesofindia.com. *Heat wave sweeps across Punjab, Haryana.* https://timesofindia.indiatimes.com/india/heat-wave-sweeps-across-punjab-haryana/articleshow/64261124.cms (2018).

21. Nasim, W. *et al*. Future risk assessment by estimating historical heat wave trends with projected heat accumulation using SimCLIM climate model in Pakistan. *Atmos. Res.* **205**, 118–133 (2018).

22. Wassmann, R. *et al*. Regional vulnerability of climate change impacts on Asian rice production and scope for adaptation. *Adv. Agron.* **102**, 91–133 (2009).

23. Carbonbrief. Mapped: How every part of the world has warmed—and could continue to warm. *Carbonbrief.org*. https://www.carbonbrief.org/mapped-how-every-part-of-the-world-has-warmed-and-could-continue-to-warm (2018).

24. Peng, S. *et al*. Rice yields decline with higher night temperature from global warming. *Proc. Natl. Acad. Sci.* **101**, 9971–9975 (2004).

25. Mahajan, G., Singh, S. & Chauhan, B. S. Impact of climate change on weeds in the rice-wheat cropping system. *Curr. Sci.* **102**, 1254–1255 (2012).

26. IRRI. *Rice Knowledge Bank—Stem Borer*. http://www.knowledgebank.irri.org/training/fact-sheets/pest-management/insects/item/stem-borer (2018).

27. Muralidharan, K. & Pasalu, I. Assessments of crop losses in rice ecosystems due to stem borer damage (Lepidoptera: Pyralidae). *Crop Prot.* **25**, 409–417 (2006).

28. IRRI. *Rice Knowledge Bank—Leafhopper* http://www.knowledgebank.irri.org/training/fact-sheets/pest-management/insects/item/planthopper (2018).

29. Krishnaiah, K. & Varma, N. Changing insect pest scenario in the rice ecosystem—A national perspective. *Directorate of Rice Research Rajendranagar, Hyderabad* **2012**, 2–8 (2012).

30. IRRI. *Rice Knowledge Bank—Blast (Leaf and Collar)*. http://www.knowledgebank.irri.org/training/fact-sheets/pest-management/diseases/item/blast-leaf-collar (2018).

31. Kobayashi, T. *et al*. Effects of elevated atmospheric CO_2 concentration on the infection of rice blast and sheath blight. *Phytopathology* **96**, 425–431 (2006).

32. IRRI. *Rice Knowledge Bank—How to Manage Pests and Diseases*. http://www.knowledgebank.irri.org/step-by-step-production/growth/pests-and-diseases (2018).

33. Prithwiraj, D. Smart water management in rice by IRRI's cost effective pani-pipe method for sustainable and climate smart rice development. *Int. J. Agri. Sci.* **9**, 4154–4155. https://bioinfopublication.org/files/articles/9_17_7_IJAS.pdf (2017).

34. Richards, M. & Sander, B. O. *Alternate Wetting and Drying in Irrigated Rice*. CGIAR Climate Change, Agriculture & Food Security—Practice Brief Climate-Smart Agriculture. https://cgspace.cgiar.org/bitstream/handle/10568/35402/info-note_CCAFS_AWD_final_A4.pdf?sequence=9&isAllowed=y (2014).

35. Zhang, H., Xue, Y., Wang, Z., Yang, J. & Zhang, J. An alternate wetting and moderate soil drying regime improves root and shoot growth in rice. *Crop Sci.* **49**, 2246–2260 (2009).
36. Terashima, K., Taniguchi, T., Ogiwara, H. & Umemoto, T. Effect of field drainage on root lodging tolerance in direct-sown rice in flooded paddy field. *Plant Prod. Sci.* **6**, 255–261 (2003).
37. IRRI. *Rice Knowledge Bank—Off-field Rice Straw Management.* http://www.knowledgebank.irri.org/step-by-step-production/postharvest/rice-by-products/rice-straw/off-field-rice-straw-management (2018).
38. IRRI. *Rice Knowledge Bank—Leaf Color Chart.* http://www.knowledgebank.irri.org/step-by-step-production/growth/soil-fertility/leaf-color-chart (2018).
39. Buresh, R. J. *Workshop on 'Balanced Fertilization for Optimizing Plant Nutrition' Sponsored by the Arab Fertilizer Association (AFA), the International Potash Institute (IPI) and the World Phosphate Institute (IMPHOS).*
40. IRRI. *Golden Rice.* https://irri.org/golden-rice (2018).
41. Sapkota, T. B. *et al.* Reducing global warming potential through sustainable intensification of Basmati rice-wheat systems in India. *Sustainability* **9**, 1044 (2017).

CHAPTER 11

Climate-Smart Maize

Abstract In the US alone, over 85 million maize tortillas are eaten each year in everything from wraps and sandwiches to pizzas and lasagnes, and that's before all our snacking on tortilla chips is added in. While Central America is its birthplace, the US, China and Brazil are now the big maize producers—together they produce two-thirds of the global harvest. One snack bag of tortilla chips has a climate footprint of around 50 grams of greenhouse gas emissions. Even with transatlantic shipping, this footprint largely comes from growing the maize in the first place. In the UK each year we throw away an estimated 23,000 tonnes of savoury snacks, including tortilla chips. Water, too much of it and especially not enough of it, embodies the climate threat to maize production. Attack by pests and diseases may increase, with a particular concern being a rise in fungal toxins, like aflatoxin, in human and livestock food. Improved plant health through soil water management, irrigation and new varieties can each give greater resilience. Supporting maize types and cultivation practices that are specifically aligned with local contexts emerges as a core requirement of climate-smart practice.

Keywords Corn • Mexico • Chiapas • Tortillas • Masa • Landraces • Drought • El Nino • GM • Low-till • Soil organic matter • Aflatoxin • Stem borer • Corn borer • Irrigation • Disease resistance

135

To go with my re-heated curry lunch is a crunchy bag of salted tortilla chips. Tortillas are made from maize flour and, as with so many great foods, we have Mexico to thank for these moreish snacks. The gigantic expanses of maize (or corn) grown around the world today are thought to owe their existence to a wild grass called Teosinte. Some 9,000 years ago, our ancestors in Central America collected this plant and over many generations domesticated and bred it into the bountiful cob-bearing plants we know and love today. Much of its journey from unassuming wild grass with small, hard seed cases to global giver of sweet, plump corn has been revealed using a wizened 5,000-year-old discovery. This incredibly well-preserved ancestral corn cob was found in the 1960s and yielded DNA that showed the transformation from wild grass to maize was already well underway five millennia ago. Though the cob was much smaller and had only half the rows of kernels (the regimented pea-sized fruits of maize) common to modern plants, it carried the unmistakable genetic signposts to softer, sweeter corn [1].

Today, maize—also called corn and going by the scientific name *Zea mays*—is the single biggest crop in the world. Rice may be the source of nourishment for over 1.5 billion people, and wheat may cover more acres but, for sheer global tonnage, maize wins hands down. It is grown on every continent save Antarctica with annual production of over 800 million tonnes. In the US alone, over 85 million maize tortillas are eaten each year in everything from wraps and sandwiches to pizzas and lasagnes, and that's before all our snacking on tortilla chips is added in [2]. While Central America is its birthplace, the US, China and Brazil are now the big maize producers—together they produce two-thirds of the global harvest [3] (Fig. 11.1).

Maize doesn't just provide the corn on the cob beloved of barbecues or the flour for our tortilla wraps and triangular snacks. It is widely used as an animal feed and processed to make everything from sweeteners and oils to drinks and glues. By its conversion to ethanol, it is the feedstock for huge amounts of vehicle fuel too. In the US, production of corn ethanol has exploded in recent decades and now accounts for almost half of the harvest. Corn ethanol can be blended with gasoline and has been promoted as a way to reduce reliance on oil imports and so increase fuel security. In theory it could also mean lower greenhouse gas emissions by substituting fossil fuels with low carbon biofuel ethanol. As we will see, however, such a field-grown answer to high carbon transport might not be all it is cracked up to be.

Maize production, 2014
Annual maize production, measured in tonnes per year.

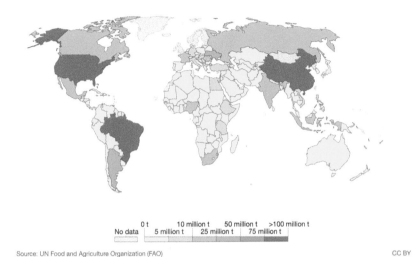

Source: UN Food and Agriculture Organization (FAO)   CC BY

Fig. 11.1 Global maize production in 2014 by country of origin (Source: Hannah Ritchie, Our World in Data) [4]. Available at: https://ourworldindata. org/grapher/maize-production

The tightly packed kernels of maize can come in a range of different forms and colours. There is sweetcorn and popcorn, the golden yellow type common in the US and Europe, and the white maize popular in many parts of Africa. It even comes in red, blue and black [5]. Our own salty lunchtime treat began life as corn grown in the fields of Chiapas State in the far south of Mexico [6]. Many centuries may have passed since the first cobs were harvested in those sun-drenched fields, but the process of turning them into crunchy snacks has changed little.

The main crop of maize in Mexico is sown in late spring and summer and then harvested over autumn and winter [7]. Waterlogged soils pose a risk during the early stages as they can mean the seedlings rot before they emerge from the ground. Maize plants like it warm—ideally an average daily temperature of at least 19 degrees Celsius and with summertime temperatures over 23 degrees Celsius—so cold weather may also be a problem. The young plants are especially susceptible to frost damage and even short periods of chill will slow growth and increase the chances of fungal attack.

After about ten weeks the plants should have grown into a towering stand 12 feet or more in height. The kernels begin to develop and the cobs to swell. It is here that dry conditions and water stress can be dangerous, as they can mean pollination is poor and development of the cobs is delayed. Just four days of water stress at this critical time can cut eventual yields in half [8, 9]. Severe hailstorms or hard frosts are even more disastrous, stripping the leaves from the plants or rupturing plant cells with ice crystals. Either risk destroying the whole crop in one fell swoop.

Finally, at around six months after sowing, the leaves on the now heavily laden plants will brown and the precious corn cobs will hopefully be ripe for harvest [9]. Frost is now no longer a risk, but lodging (falling over) of plants due to disease, pests or severe weather is. The sweet kernels need to have dried to the right moisture content (around 30 per cent) before they are harvested to avoid mould growth, and even then, they need to be further dried if they are to be stored for an extended period.

After harvest the corn destined to be our premium tortilla chips must first be made into a coarse dough or masa. A mixture of corn, water and food-grade lime (to remove the skins on the corn kernels and give a longer shelf life) is cooked up together in large steam kettles while being constantly agitated through injections of steam and compressed air. After cooking in this lime solution (called nixtamalisation), the mixture is rapidly cooled and allowed to steep to get rid of the skins and to soften the kernels. A thorough wash and drain, and then it's into the grinding room where heavy stones made from hard volcanic rock mash the mixture into the coarse masa that is needed. If destined for foreign markets, like the snack manufacturers of Europe, it is dried and graded into flour before being packaged up and sent by ship across the Atlantic. On arrival the flour is rewetted into dough and the raw chips are rolled and cut before being baked, cooled, and fried and seasoned. Our snappy tortilla chips are then ready for eating [10].

One snack bag of tortilla chips has a climate footprint of around 50 grams of greenhouse gas emissions. Even with transatlantic shipping, this footprint largely comes from growing the maize in the first place. Emissions of nitrous oxide from adding fertilisers to the soils again dominate, making up around one-third of the total. Some also occur via the energy used in making fertilisers, pesticides and herbicides, and from operating farm machinery, artificial irrigation or corn drying equipment [11]. Once harvested, the electricity and gas used for processing the corn into masa are important sources of emissions, with oil for frying, packaging and of course transport each adding to the overall total.

Our appetite for savoury snacks like tortilla chips seems boundless. Global sales are predicted to top £100 billion by 2020, with the average person in the UK already tucking away around 7 kilograms of snacks a year, rising to a thirst-inducing 9.5 kilograms per person each year in the US [12]. Issues of dietary health and expanding waistlines aside, meeting such rising demand for maize and other crops will mean increasing greenhouse gas emissions unless global production becomes more climate-smart. Alongside the hidden 'wastage' that is overconsumption of food comes the more overt waste of tortillas and maize snacks that means huge amounts of land cultivation and emissions for no ultimate benefit. It should by now be a familiar story.

As the tortilla chips we buy from the shops are usually well wrapped and preserved, they can have a good long shelf life. Like many snack foods, however, once opened, the race to beat staleness and waste begins. In the UK each year we throw away an estimated 23,000 tonnes of savoury snacks, including tortilla chips [13]. Almost all of this is deemed avoidable and mostly results from the snacks not being eaten in time or from too much being served. While still some way behind potato crisps as Britain's favourite snack, tortilla chips have become a mainstay of the weekly shopping basket and represent a global business worth over $10 billion [14]. As with all the foods and drinks we've followed so far, avoiding wastage has the potential to save thousands of tonnes of emissions. For our tasty maize-based snacks, buying and serving only what is really needed, storing them in air-tight packets once opened and keeping an eye on use-by dates will all go a long way to realising this potential.

* * *

Water, too much of it and especially not enough of it, embodies the climate threat to Mexican maize production. The drought conditions caused by the El Nino of 1998 knocked a quarter of the annual harvest, while the wetter more humid years of 1991 to 1993 provided bumper crops [15]. The region of Chiapas and its indigenous farmers in southern Mexico are no strangers to such variability. For generations they have had to cope with the vagaries of El Nino and La Nina—those fluctuating states of warm and cold waters in the Pacific Ocean that drive big swings in temperature and even bigger swings in rainfall patterns over large swathes of the world.

Traditionally Chiapas farmers have grown maize alongside beans, using their crops to feed themselves and their families rather than the international snack trade. Maize occupies half of all the land used for agriculture in the region and is worth an estimated $400 million each year. The individual farms are often very small though—many have been divided and sub-divided several times as they are passed down from one generation to the next. They can also be very poor. Per hectare, average maize yields in Chiapas are two-thirds that of Mexico as a whole with access to financial and technical support limited and only 1 farmer in 25 using irrigation [16].

Drought has been a common feature of the past decade. Severe drought gripped much of Mexico during both 2011 and 2012, with maize farmers in the north being particularly badly hit and the national maize harvest falling by several million tonnes [17]. In the powerful El Nino year of 2016, Chiapas itself endured such an extreme drought that water levels in its Nezahualcóyotl reservoir dropped by 25 metres and a sixteenth century church not seen for a generation emerged [18]. With so few farms having access to irrigation, any major decrease in rainfall due to climate change ramps up the risk of widespread crop losses. As early as 2030 annual rain-fall in the region is predicted to decrease by between 6 and 53 mm, along-side average temperature increases of over 1.5 degrees Celsius [16].

For the 20 million people in Mexico who rely on rain-fed maize as their staple food, this trend towards drier conditions through the century strikes at the very heart of food security. Just how big the impacts will be depends to a large extent on which emissions pathway global society decides to take. With deep and rapid actions that limit global average temperature increase to around 2 degrees Celsius, some areas of Mexico—such as central and northwest states—could see wetter conditions and higher yields. However with further growth in global emissions, and warming of 4 degrees Celsius or more [19], most states are predicted to experience much drier conditions and precipitous drops in maize yields. For some, including Chiapas, it would spell disaster—yields there could drop by a famine-inducing 80 per cent [15].

Damage from tropical cyclones may also increase as higher sea surface temperatures enhance the intensity of storms. Likewise, more extreme rainfall events and persistent flooding of fields bring with them the threat of complete loss of crops as seedlings and young plants rot in the waterlogged soils [16]. Even where such severe weather impacts do not destroy crops directly, they may open the plants up to more attack from pests and diseases. Climate change, it seems, is shaping up to give the world's biggest crop some very hard falls.

Between 1988 and 1990, an estimated $44 billion-worth of maize was produced around the world, but a further $27 billion-worth was lost due to disease, pests and weeds [8]. In North America, there is now concern that the ranges of the most damaging maize pests—the corn borer, the corn earworm and the corn rootworm—may all expand as winters become milder [20]. An inspired response to another of these rampant maize pests (the stem borer) has been one championed by smallholders in Uganda. It's called push-pull.

This clever combination of companion planting with maize involves growing stands of tall Napier grass around the plots of maize. In amongst the maize itself, a low-growing nitrogen-fixing plant called *Desmodium* is also grown. The *Desmodium* enriches the soil and suppresses weeds while releasing a chemical that repels the stem borer moths (the push). Meanwhile the Napier grass releases a chemical that attracts the moths (the pull). When the stem borer moths lay their eggs on the Napier grass, the grubs that hatch are quickly smothered by the strong sap of the grasses while the maize crop can grow on free of the stem borer. Using this system not only boosts maize yields but also provides three crops instead of one—the Napier grass and the *Desmodium* both provide good fodder for livestock [21].

Like some maize pests, fungal diseases may benefit from a changing climate too. Warmer and more humid conditions allow faster reproduction and spread of spores. As maize plants reach maturity, any waterlogging of the soils can provide ideal growing conditions for damaging fungal diseases such as the delightfully named 'crazy top' and 'common smut'. Another, called maize ear rot, thrives in the hotter drier conditions predicted for much of Mexico [22].

Not only do fungal attacks put at risk the size of the crop, they can also endanger the lives of anyone that then eats it. Some fungi naturally produce toxins as they grow on fruits, nuts, and cereal grains [23]. These mycotoxins can contaminate the crop in the field or during storage and transport after harvest. Globally they are responsible for numerous deaths and incidences of disease, including liver cancer and immune suppression [24].

One type that commonly affects maize is called aflatoxin, and it is very dangerous indeed. Aflatoxin is one of the most potent naturally occurring liver carcinogens known and, in large doses, it is deadly [24]. Initial symptoms of poisoning include abdominal pain and vomiting, with chronic exposure associated with stunted growth in children. An outbreak in Kenya in early 2004 killed 125 of the 317 people affected. Contaminated maize was identified as the source, with many of the affected households having stored wet maize in their homes and so increased the risk of fungal growth [23].

Another group of mycotoxins common to maize are the fumonisins. These are produced by *Fusarium* fungi and have been linked to oesophageal cancer and birth defects in humans, and to liver and kidney toxicity in animals. Because of the great danger it poses, mycotoxin contamination is strictly regulated in much of the world. Grain harvests are regularly screened to ensure they are within safe toxin levels and, if not, are used for non-human consumption or destroyed. In the US this destruction of contaminated food costs over $1 billion a year, with maize farmers bearing the brunt of losses.

In Mexico too there are defined limits on aflatoxin contamination levels—a low one for maize destined for human consumption and higher limits for that used as livestock feed. In 1989, the testing of the maize in the northeastern state of Tamaulipas revealed that almost the entire 60 thousand tonne harvest had aflatoxin concentrations far above the level permitted for human consumption. Following many failed attempts to remove the toxins, the contaminated maize was instead used to make alcohol. Down in Chiapas there has been concern that the drinking of Pozol—a foamy fermentation of maize popular with indigenous inhabitants—may also pose aflatoxin dangers. Thankfully, the testing of Pozol from over 100 local markets found only 1 to have concentrations above the recommended level [25].

How exactly a future climate will change mycotoxin risks in Mexico remains unclear. Higher temperatures favour infection by fungi like *Aspergillus flavus* that are known to produce aflatoxins [24]. Warmer winters may also promote maize pests, such as the dusky sap beetles of the US corn belt, that carry fungal diseases from plant to plant. Certainly the warning for the US is that mycotoxin concentrations in maize are set to increase. Down in Mexico the maize farmers and Pozol drinkers of Chiapas could well be at even greater risk.

*   *   *

With drought being the prime climate change threat for so many Mexican maize farmers, a transition from rain-fed to irrigated farming seems an obvious response. In southern states like Chiapas, over two million hectares of usable farm land sits uncultivated through the dry autumn and winter months, and the Mexican Government has itself highlighted the urgent need for investment to upgrade irrigation systems and embed resilience [26]. Irrigation could allow unused lands to become productive farms, could expand the number of harvests in dry areas from one into two each year and might provide growers with the vital bridge they need to span the expanding periods between one wet season and the next.

The potential is huge. Experiments using fully irrigated maize in Chiapas have delivered yields of up to ten tonnes per hectare—five times the level commonly achieved by smallholders at the moment. The task of ensuring access to irrigation for the tens of thousands of rain-fed maize farmers across southern Mexico is similarly gargantuan. Electricity supplies would need to be extended to power a new network of water pumps, reservoirs would need to be extended, and the expanding draw on water resources in the South would need to be balanced with growing water demands further north as drought conditions begin to bite there too. Across the whole country the efficiency with which water is used will need to improve radically—in some districts that already use irrigation, the water use efficiency currently sits at a low and very leaky 37 per cent.

Even in the absence of irrigation, there are strategies that can help. Around one-quarter of farmers have taken to planting leguminous cover crops—nitrogen-fixing crops like beans that are grown in rotation with the maize, reduce soil erosion in heavy rains, and can be ploughed into the soil to give a boost to its fertility [16]. More than half of maize growers have adopted 'minimum tillage', where the reduced soil disturbance helps the retention of organic matter and water below the surface. This approach has been successful on larger farms with flat land and access to specialised planting machinery, but it could compound problems on smaller farms. In southern Mexico an array of native breeds adapted for local conditions (called 'landraces') are commonly grown, and these often have weaker roots than the new commercial varieties. Minimum tillage makes it harder to plant seeds deep enough and can also mean more rain runs off the surface of sloping fields. For those without planting machinery or with hillside farms, it therefore runs the risk of increasing climate vulnerability instead of reducing it.

Encouraging smallholders to switch away from traditional maize varieties to higher-yielding hybrids has long been seen as the way in which Mexico could boost harvests—especially if combined with greater levels of access to fertilisers, irrigation and machinery. Despite large investments, the success of such programmes has been limited. Many farmers appear reluctant to give up their native landraces. They may well be right. The locally adapted characteristics of native plants could provide more resilience to future drought, flood and pest risks than some of the new 'high-yield' varieties.

Genetically modified or transgenic maize has been touted as a way to ensure both higher yields and the much-needed resilience in a future climate, allowing deliberate selection of traits like pest and disease resistance [27]. In theory, transgenic maize could be developed for an optimal fit with current and future growing conditions. In reality, those conditions are locally specific and widespread replacement of native varieties with just one or two transgenic crops risks losing the very diversity that is the backbone of resilience.

Instead of GM varieties and the broad-brush application of higher-yield hybrids, a more nuanced approach to boosting harvests in a changing climate has been suggested. Here, the many native landraces (over 50 have been identified) and the myriad attributes they have developed for success in local contexts are made full use of [28]. By identifying features in specific landraces that could enhance climate change resilience in a certain area, and then ensuring farmers there have access to it, truly climate-smart maize production is possible. Such a system requires detailed climate projections, seed banks and plant breeding facilities [29], alongside an established framework for farmer interaction and consultation. As ever, strong and well-funded extension services would be crucial in understanding local contexts and providing access to the right seeds at the right times [28]. There is already precedent for this.

Mexico's 'PROEMAR' project delivered impressive gains in maize yields for the farmers involved. Even in poor-weather years harvests ballooned by over 50 per cent and average output per hectare hit more than eight tonnes. This programme concentrated on improving extension services for smallholders, including soil testing and assistance with precision fertiliser application. Farmers were trained in good practice for seed treatment and sowing, improved planting densities, and how best to balance fertilisers with the needs of the plants as they grew (helping to reduce costs, nitrous oxide emissions and pollution of drainage waters) [26]. Such integrated approaches can extend well beyond the maize fields themselves. To address the threat posed by post-harvest spoilage, and especially that of mycotoxins, more climate-resilient grain storage is now being championed alongside improvements to supply chains and access to markets [30].

The barriers to realising these kinds of successes for all farmers in Mexico, as the twenty-first century progresses, are formidable. The UN's Food and Agriculture Organisation identified low productivity, low organisational capacity in farmers' organisations and poor access to financial products as the central ones that must be overcome [16]. They highlight the need for agricultural insurance (to protect livelihoods in bad years), early warning systems (to allow proactive responses to severe weather events like droughts), and an increase in the overall awareness of climate-smart technology and management options. Investment in capacity and support can deliver big returns—PROEMAR used $1.7 million to provide income gains of $9.3 million for its farmers.

There is a popular saying in the maize-growing heartlands of Mexico: '¡Sin Maíz No Hay País!'—'Without Corn There is No Country!' [31]. Climate change poses an existential threat to the maize farmers of Mexico. With sufficient support, climate-smart strategies could help safeguard this staple food source for millions and underpin the future of the country itself.

REFERENCES

1. Briggs, H. Ancient corn cob shows how maize conquered the world. *BBC Online*. https://www.bbc.co.uk/news/science-environment-37999506 (2016).
2. Weber, R. J. *Shelf Life Extension of Corn Tortillas*. Masters dissertation, Kansas State University. https://core.ac.uk/download/pdf/5165140.pdf (2008).
3. Ranum, P., Peña-Rosas, J. P. & Garcia-Casal, M. N. Global maize production, utilization, and consumption. *Ann. N. Y. Acad. Sci.* **1312**, 105–112 (2014).
4. Ritchie, H. Global maize production, 2014. *Ourworldindata.org*. https://ourworldindata.org/grapher/maize-production (2018).
5. O'Leary, M. Maize: From Mexico to the world. *CIMMYT.org*. https://www.cimmyt.org/maize-from-mexico-to-the-world/ (2016).
6. Sweeney, S., Steigerwald, D. G., Davenport, F. & Eakin, H. Mexican maize production: Evolving organizational and spatial structures since 1980. *Appl. Geogr.* **39**, 78–92 (2013).
7. FAO. *Mexico. GIEWS—Global Information and Early Warning System*. http://www.fao.org/giews/countrybrief/country.jsp?code=MEX (2018).
8. Rosenzweig, C., Iglesias, A., Yang, X., Epstein, P. R. & Chivian, E. Climate change and extreme weather events; implications for food production, plant diseases, and pests. *Glob. Chang. Hum. Health* **2**, 90–104 (2001).

9. Darby, H. & Lauer, J. *Critical Stages in the Life of a Corn Plant.* University of Wisconsin Extension. http://corn.agronomy.wisc.edu/Management/pdfs/CriticalStages.pdf (2018).

10. madehow.com. *Tortilla Chip.* http://www.madehow.com/Volume-1/Tortilla-Chip.html (2018).

11. Grant, T. & Beer, T. Life cycle assessment of greenhouse gas emissions from irrigated maize and their significance in the value chain. *Aust. J. Exp. Agri.* **48**, 375–381 (2008).

12. Stones, M. Global savoury snacks sales to hit £103bn by 2020. *foodmanufacture.com.* https://www.foodmanufacture.co.uk/Article/2016/09/08/Global-snack-food-market-valued-at-103-bn-by-2020 (2016).

13. WRAP. Household food and drink waste in the United Kingdom 2012. *Waste and Resource Action Programme.* http://www.wrap.org.uk/sites/files/wrap/hhfdw-2012-main.pdf.pdf (2013).

14. Euromonitor.com. *What's New in Sweet and Savoury Snacks.* https://blog.euromonitor.com/whats-new-in-sweet-and-savoury-snacks-opportunities-abound-for-a-new-wave-of-products/ (2015).

15. Murray-Tortarolo, G. N., Jaramillo, V. J. & Larsen, J. Food security and climate change: The case of rainfed maize production in Mexico. *Agri. For. Meteorol.* **253**, 124–131 (2018).

16. CIAT. Climate-Smart Agriculture in Chiapas, Mexico. *CSA Country Profiles for Latin America Series.* Washington, DC: The World Bank Group. http://sdwebx.worldbank.org/climateportal/doc/agricultureProfiles/CSA-in-Chiapas-Mexico.pdf (2014).

17. Torres, N. Mexican farmers suffer worst drought in 70 years. *Reuters.* https://www.reuters.com/article/us-mexico-drought/mexican-farmers-suffer-worst-drought-in-70-years-idUSTRE7AO18Q20111126 (2011).

18. UNISDR. *Drought-hit Chiapas Leads on Sendai.* https://www.unisdr.org/archive/49048 (2016).

19. Carbonbrief. Mapped: How every part of the world has warmed—And could continue to warm. *Carbonbrief.org.* https://www.carbonbrief.org/mapped-how-every-part-of-the-world-has-warmed-and-could-continue-to-warm (2018).

20. Diffenbaugh, N. S., Krupke, C. H., White, M. A. & Alexander, C. E. Global warming presents new challenges for maize pest management. *Environ. Res. Lett.* **3**, 044007 (2008).

21. Pickett, J. A., Woodcock, C. M., Midega, C. A. & Khan, Z. R. Push–pull farming systems. *Curr. Opin. Biotechnol.* **26**, 125–132 (2014).

22. Doohan, F., Brennan, J. & Cooke, B. *Epidemiology of Mycotoxin Producing Fungi* 755–768 (Springer, 2003).

23. WHO. *Mycotoxins.* http://www.who.int/news-room/fact-sheets/detail/mycotoxins (2018).

24. Wu, F. *et al.* Climate change impacts on mycotoxin risks in US maize. *World Mycotoxin J.* **4**, 79–93 (2011).
25. García, S. & Heredia, N. Mycotoxins in Mexico: Epidemiology, management, and control strategies. *Mycopathologia* **162**, 255–264 (2006).
26. Fernández, A. T., Wise, T. A. & Garvey, E. *Achieving Mexico's Maize Potential* (Tufts University, 2012).
27. Dalton, R. Mexico's transgenic maize under fire. *Nature* **462**, 404 (2009). https://www.nature.com/news/2009/091125/full/462404a.html.
28. Hellin, J., Bellon, M. R. & Hearne, S. J. Maize landraces and adaptation to climate change in Mexico. *J. Crop Improv.* 28, 484–501 (2014).
29. O'Leary, M. Maize: From Mexico to the world. *cimmyt.org.* https://www.cimmyt.org/maize-from-mexico-to-the-world/ (2016).
30. FAO. *Modernizing for Growth: The Case of Grain Storage in Mexico.* http://www.fao.org/in-action/agronoticias/detail/en/c/1118464/ (2018).
31. McCune, N. M. *et al. Sustainable Development-Authoritative and Leading Edge Content for Environmental Management* (InTech, 2012).

PART IV

Dinner

CHAPTER 12

Climate-Smart Potatoes

Abstract For a long time, Europe, North America and the former Soviet Union were the powerhouses of world potato growing. Since the 1960s, though, production in Asia, Africa and Latin America has more than quadrupled, and China and India between them now grow over one-third of the enormous 350 million tonne global potato harvest. Every tonne of raw chips has a carbon footprint of just under a tonne of greenhouse gas emissions. Each year in the UK, we discard some 320,000 tonnes that could have been eaten. This veritable mountain of dumped potatoes represents an annual climate penalty of well over 80,000 tonnes of greenhouse gas emissions. Drought is a major risk for many growers, as few in the UK use irrigation, and the viable area of rain-fed potatoes could shrink to 5 per cent of its current extent as droughts intensify in twenty-first-century Britain. Diseases such as late blight also pose a big threat for growers around the world. A combination of disease and drought-resistant varieties, along with irrigation, soil management and greater farm nutrient efficiency can deliver much greater resilience and more secure yields, while driving down emissions.

Keywords Scotland • Maris Piper • Irrigation • Blight • Scab • Drought • Waterlogging • Chips • French fries • Field hygiene • Seed potatoes • Cool storage • Bruising

151

For dinner it's a lip-smacking indulgence enjoyed in homes right across Britain: fish & chips. Voted as the nation's number one takeaway meal, a quarter of a billion chip shop meals are sold each year. Its joy has stayed with me since that very first newspaper-wrapped parcel of delight, eaten while huddled in a beach towel after a day of icy paddling in the North Sea. Britain's first chips reputedly went on sale in 1850s Yorkshire thanks to a lady called Granny Duce. They quickly became a hit here in Scotland too. Today over a third of Scots eat chips two or more times a week [1] and across the UK we chomp our way through one million tonnes of them each year [2].

The humble chip-yielding potato has come a long way from the high hills of the tropical Andes, where its wild ancestors were first domesticated over 5,000 years ago [3]. Introduced to Europe by the Spanish in the sixteenth century and helping power global population growth throughout the 18th and 19th centuries [4], it is now the world's fourth most important food crop after maize, wheat and rice.

For a long time Europe, North America and the former Soviet Union were the powerhouses of world potato growing. Since the 1960s though, love for growing and eating this versatile tuber has spread right around the world. Production in Asia, Africa and Latin America has more than quadrupled, and China and India between them now grow over one-third of the enormous 350 million tonne global potato harvest [5] (Fig. 12.1).

To fit the myriad climate envelopes and culinary preferences of a global audience, the potato plant has undergone centuries of selective breeding. The UK's Agriculture & Horticulture Development Board lists no fewer than 333 different varieties spanning the full spud-you-like alphabet from Accent to Zohar and including intriguing options like the Moulin Rouge— 'unusual long tuber, good blemish resistance'—or the Picasso—'pink eyes and creamy flesh, resistant to common scab' [7]. Across in their native Andean home, work is ongoing to conserve and learn from the rich heritage of varieties and growing practices there [8].

In more temperate climates like that of Northern Europe, most potatoes enjoy a mean daily temperature of around 18 degrees Celsius, plus night-time temperatures that drop under 15 degrees Celsius to trigger the formation of the tubers. If it gets too cold (below 10 degrees Celsius) or too hot (above 30 degrees Celsius), tuber formation will slow or stop altogether. In the tropics, varieties are grown that are better able to cope with high temperatures and that can do well even with their shorter day-lengths compared to the long summer days of higher latitudes [3].

Potato production, 2014
Annual agricultural production of potatoes, measured in tonnes per year.

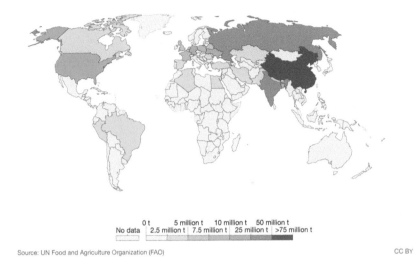

No data	0 t	5 million t	10 million t	50 million t
	2.5 million t	7.5 million t	25 million t	>75 million t

Source: UN Food and Agriculture Organization (FAO)   CC BY

Fig. 12.1 Global potato production in 2014 by country of origin (Source: Hannah Ritchie, Our World in Data) [6]. Available at: https://ourworldindata. org/grapher/potato-production

Potatoes like well-drained and aerated soil—in much of the world it is common to grow them in mounds or ridges to ensure their roots don't get waterlogged. But they are thirsty plants too. A shortage of water as potatoes form results in many being deformed and spindly. They may also get very scabby. So-called common scab is a problem for many farmers without access to irrigation. It often appears where the surface of the developing potato has dried out, producing unsightly brown pits and ridges [9]. Frequent watering during dry periods is the best scab-avoidance strategy, but leaving it too late and then dousing the fields can induce even bigger problems. If the plants become too water stressed and then receive lots of irrigation water, the potato tubers can split, opening them up to infection and leaving a nasty surprise when it comes to harvest in the form of a field of potatoes containing rotten black hearts [3].

Our own chip shop potatoes are good old Maris Piper, described as being 'high yielding, resistant to gangrene, and good for cooking and frying'. They are grown in the next county along from us and so have only

40 miles or so to travel from field to deep fat fryer. In Scotland it takes around three months from the planting of seed potatoes to harvest and, with willing weather, two or three crops a year are possible—the earlies, the main crop, and then perhaps a late planting to give potatoes in time for Christmas dinner. Seed potatoes—tubers grown specifically for re-planting and usually certified as disease free—are planted a few inches deep and begin sprouting up into daylight about a fortnight later. With the right combination of water, light and temperature, the plants soon generate multiple stems, spread their roots and begin to form new tubers. This is a crucial time for the size and quality of the eventual crop. Any brake on the development of the tubers, even short drought periods, means the final harvest will be smaller [10].

Within two months of planting the tubers should be filling out, a few weeks later the mature potato crop can be dug up. The relatively large weight and size of potatoes means damage during harvesting, cleaning and grading is a common problem. Scottish potato farmers have a specific guide on how to keep their harvest safer as it passes from soil to trailer, and from storage shed to delivery truck [11]. Evoking images of lab coats and large mallets, the guide even includes a damage league table, giving bruising and shatter risk scores for each of the main varieties.

Having survived the threats of field and farm machinery, our potatoes are almost ready for the fryer. The local fish and chip shop peels and chips the potatoes before deep frying in oil and handing them over wrapped in paper and smelling utterly delicious. In terms of carbon footprints, fresh potatoes are relatively light on their feet. They can rack up substantial emissions through the energy used to store them though, as they often need to be refrigerated in the summer months. Other emissions arise from the production of the seed potatoes, use of fertilisers and pesticides, and energy used for irrigation [12]. Overall, the growing, transport, storage and processing add up to just under tonne of greenhouse gas emissions for every tonne of raw chips that arrives at the takeaway [13] (or around 250 grams for a good mealtime portion [14]). But that's before they are cooked. Frying our chips, whether at home or in a shop, uses a large amount of oil and energy and represents the biggest component of a chip's life-cycle footprint. Using the average commercial deep fat fryer, the carbon footprint of our single portion of chips is doubled to around half a kilogram [13]—more if your local chippy uses palm oil, less if they use sunflower or rapeseed oil [15]. A sprinkle of salt and a dash of vinegar won't do much to change this, though if you have a penchant for smothering your chips in ketchup, this will bump up the footprint by another 15 grams or so [16]. This may all look bad in the carbon stakes, but there is a big climate-saving grace for chips compared to most potatoes: we're more likely to eat them.

Any city high street on a Friday night or gull-circled seaside promenade can testify to the fact that we throw away chips. Compared to the huge mass of other potato meals that are wasted, however, our chips do pretty well. Each year in the UK, we discard over 700,000 tonnes of potatoes—equivalent to six million spuds every day and second only to bread as the nation's biggest food-bin filler. Much of this waste is in the form of peel and deemed 'possibly' avoidable, but some 320,000 tonnes are definitely avoidable. Not being used in time is again the most commonly cited reason for this waste—rare is the British grocery cupboard in which at least a couple of green and sprouting potatoes can't be found. Many are wasted because too much is cooked or served, with the remainder being wasted due to personal preference or accidents like burning the dinner [17]. This veritable mountain of dumped potatoes represents an annual climate penalty of well over 80,000 tonnes of greenhouse gas emissions.

The now hackneyed advice to cut such wastage by keeping track of what we have and only buying and cooking what we need still applies. Storing potatoes well—in dark, cool, well-ventilated conditions—can also prolong their usable lives and prevent them sprouting. Even where they have developed green patches or sprouts, these can be cut off and the rest used [18]. Cooking and serving only what is needed is easier said than done in most households, but the recipes for re-using leftover potatoes are legion [19]—leftover chips crammed between two slices of buttered bread have a special place in heaven.

* * *

Britain is generally an ideal place to grow potatoes. We have the right temperatures, soils and annual rainfall to produce bumper crops. Scotland boasts some of the world's best seed potato producers and should be well placed to reap the spud-swelling benefits of a carbon dioxide-enriched atmosphere too. The future of our chip supper in the face of climate change would therefore seem secure, but recent severe weather events tell a different story.

The start of 2012 in the UK was a dry one. Potato farmers across the land were able to get onto their fields early and get the first crop sown into the dry soil, but then needed rain to help the young plants along. The rains still didn't come. Our government held a drought summit to discuss the prolonged dry conditions affecting the southeast of the country. By March the area officially in drought had extended into northern England, with wild fires breaking out in Wales and Scotland [20]. The early potatoes were struggling. Drought conditions in the young vegetative stage hobble foliage and root growth, while at later stages it can mean deformed tubers and a plummet in the overall harvest [10]. Irrigation was used by some farmers, most just looked to the horizon for rain clouds.

Then, as April was drawing to a close, the skies across Britain darkened. In just a few days, some areas experienced more than three times their average rainfall for a whole month—homes were flooded, roads and rail lines cut. In Scotland the rain kept on coming. Soils that had initially been parched became sodden paddies, the uniform ridges of potato plants interspersed with long moats of standing water. Many crops of earlies simply could not be harvested. The drought conditions had already weakened the plants, now the moist air and sodden soils were the cue for damp-loving potato diseases to take their toll. Infection by black leg—a bacterial disease that dissolves the cell walls of plants—hit highs not seen in Scotland for 20 years [21]. Then came a surge in the potato's most infamous foe of all: blight.

Late blight is a highly destructive disease caused by a fungus called *Phytophthora infestans*. It first arrived on the shores of an unsuspecting Europe in the 1840s. With alarming speed it spread through the continent's farms, its tell-tale lesions on foliage and wet and dry rot of the tubers soon appearing in Ireland. The impact was devastating. Waves of the disease burned through Irish potato harvests for year after starving year. It putrefied the main food source for more than one-third of the population. An estimated 1 million people died of starvation and epidemic disease in the space of just five years. Another two million quit Ireland for foreign—and hopefully less blighted—shores like the Americas [22].

Blight likes wet conditions, with high humidity and temperatures of 15 to 20 degrees Celsius being optimal for it to grow and release its many millions of infectious spores. These spores can swim in the thin films of water on leaf surfaces [23]—they have whiplash type flagella to propel them—and once they've found their target, they quickly encyst and germinate. Each sends out a germ tube that penetrates the plant tissues. Within a few days lesions begin to appear on the infected plant's leaves or stems. Initially these are small and irregular, then expanding to form a circle of brown dead tissue. During warm and moist periods, whole plants (indeed whole fields of plants) can be blighted. Infected fields carry a distinctive dank and mildewy smell—the first warning whiff on the breeze of the putrid inedible mess that is being made of the tubers underground.

Soon after a lesion appears, the new fungus is itself ready to produce spores. A single lesion may produce 100,000 sporangia—protective capsule structures containing the spores that can themselves be swept up into the wind and travel several kilometres to infect potato pastures new [24]. Infected material, such as tubers kept in storage or left in the ground, provide overwinter disease banks, while the international shipping of seed potatoes provides it with a readymade global distribution network.

Almost all potato areas across Asia, Africa, Europe and the Americas have blight now or have had it in the recent past. Blight-free areas tend to be those that are cold or hot enough to kill it off—those at very high or low latitudes and altitudes. As of 2008 the global cost of potato blight was estimated at over $6 billion a year. A warmer future is predicted to further increase risks in cooler areas and the earlier-onset of outbreaks [25]. Nearly two centuries on from the Irish Famine, late blight still poses a major threat to the food security of millions.

Too much rain then, even for the thirsty potato plant, can pose a threat. As such, the climate change projections for eastern Scotland make for worrying reading for growers. By the middle of the century, winter rainfall may increase by one-fifth [26], with that rain falling in ever more intense downpours that risk stripping the soils of nutrients and making planting of the early potato crop an exercise in professional mud management.

Future summers are set to go in the opposite direction, becoming up to one-third drier by the 2050s, alongside a hike in maximum daily temperatures of over 4 degrees Celsius. A taste of this dusty future was delivered in 2018 when, after a late spring, the UK experienced one of its hottest and driest summers on record. In July temperatures peaked at over 35 degrees Celsius in the south and wildfires became a major issue over in Wales. Lightning storms left 30,000 homes without power, with a mini-tornado and hailstones the size of £1 coins reported. When any rain did fall, it came in torrential downpours, leaving roads blocked and drain covers blasted out of the ground [27]. On the potato farms, the combination of high temperatures and little rainfall was stunting plant growth and shrivelling the tubers. Evaporation rates from the soils in Scotland were equivalent to those a thousand miles south in central France, but only around one-third of the Scottish farmers had irrigation systems in place [28]. Across northwest Europe the harvest fell by around one-fifth, with many farmers reporting poorer sizes and quality [29]. Even the UK's popular press were alarmed, warning that chips would be an inch shorter due to the droughts [30].

For our own Maris Pipers, the impacts of climate change by 2050 are a distinctly mixed hessian bag. Higher carbon dioxide concentrations in the atmosphere and longer growing seasons might mean a slight boost in yields [31]—about 5 per cent under current practices or up to 15 per cent if all their higher demands for water and fertilisers are met. Supplying enough water is the key. The use of irrigation water would likely need to rise by a third, and this future climate of hotter, drier summers would mean water demands exceed current supplies for almost half of the farms [32]. The area suitable for traditional rain-fed potato growing—such as eastern Scotland—could shrivel to a scabby one-twentieth of its current extent as droughtiness intensifies [33]. Without irrigation, many of these potato farms would cease to be.

Globally the climate scenarios for potatoes give a similarly mixed picture of some winners and potentially huge numbers of losers. In northern India and across the highlands of South America, Africa and Asia, yields could see an uptick of one-fifth or more as plants benefit from more carbon dioxide [25]. But in large regions of North America and Eastern Europe, potato production is set to fall precipitously. As drought, disease and pest damage intensifies, some areas could see more than half their harvest wiped away. Other regions, including many farms in the northeastern US, the Caribbean and southwest Russia, risk being obliterated from the world potato-growing map altogether. Overall, a small drop in worldwide yield is predicted by the 2050s, rising to a reduction of up to one quarter towards the end of the century [34]. If realised, this would put a major tuber-shaped dent in global food security.

*   *   *

The projected impacts of climate change on potatoes may be dire for many, but most scenarios assume no change in which varieties the farmers will choose and how they will grow them. In reality, farmers will adapt to the changes they see—called autonomous adaptation—by switching varieties or altering their planting and harvesting dates.

For the spectre of drought that hangs over many farms, the installation of irrigation is an obvious route towards greater resilience. There are costs in terms of the equipment, the energy used for water pumping and the additional stress this could put on wider water demand, but if it can be coupled with on-farm rainfall capture and storage, it has the potential to be lower carbon, avoid major losses in yields and more than outweigh the upfront costs of installation. Recent years have seen a surge in the creation of winter-filled reservoirs on farms in England [31]. These have the dual advantage of helping to manage the impacts of more intense rainfall events and their soil-scouring risks, while banking the excess water for the hotter, drier summer months ahead.

As demand for irrigation water spirals upwards, more efficient use of it can also help ensure that on-farm supplies meet the thirsty needs of the potato crop without hiking up carbon emissions. Careful scheduling of watering alongside good soil monitoring and weather forecasting—to best meet the changing needs of the plants as they develop—pay big dividends. Equipment like automated drip or trickle irrigation allows growers to deliver water to exactly where and when it is needed, and to avoid some of the wastage and uneven supply issues common to rain guns and sprinklers [35]. By incorporating more organic material like composts and green manures [36], the soils themselves can be managed to increase their fertility, workability and how much moisture they hold. Likewise, encouraging the plants to develop deeper roots—by providing irrigation in large sustained drenchings rather than lots of small sprinklings—means that drying out of surface soils during a drought will do less damage [37].

For farmers without access to irrigation water, switching to more drought-resistant varieties can bring much-needed resilience as weather extremes become even more extreme. Desiree potatoes, for example, are able to respond to dry periods by diverting water and resources to their roots and tubers instead of the shoots and leaves—banking their reserves for a rainy day. Our own Maris Pipers are already a pretty good choice in terms of their ability to endure limited drought periods and then make good use of any rainfall later in the growing season. Others, like the UK's widely grown Lady Rosetta, have a much tighter growth window, and so, even a short-lived drought can be very destructive [31].

With wetter winters, timing the planting and harvesting to avoid saturated soils, together with the use of lighter and broader-wheeled machinery, can reduce soil damage and compaction. Well-maintained and operated harvesting, storage and transport systems will also cut losses due to damage and spoilage [11, 38]. For pests and diseases like late blight, improved field hygiene and use of resistant varieties can prevent serious outbreaks [39]. Wherever such strategies increase climate resilience and productivity, they are likely to give indirect emissions savings— each additional potato that makes it through the minefields of blight and drought avoids the need to grow a replacement. By using farm nutrient budgeting and greater precision in the quantity, timing and placement of fertilisers, direct emissions can be radically reduced too, while some potato varieties also produce copious amounts of shoots and leaves that can then be incorporated into the soil to boost its carbon content.

Around the world, the climate-smart options that will work best for a particular farm will depend on its local circumstances. Growing a super-resilient low carbon potato that nobody wants to eat is worse than pointless. Instead, working with growers to identify climate risks and opportunities in the context of the myriad other demands they face is more likely to deliver lasting benefits [40]. For some this will mean a transition away from potato growing and diversification into new crops. For others, it will mean an increase in fertiliser, water and pesticide use—boosting resilience and productivity at the expense of emission reductions. But where any such hike in inputs is able to induce an even bigger boost in yields, then the climate impact of producing each individual potato (the emissions intensity) is still reduced.

For the potato farmers of Ethiopia's Rift Valley, just such increased intensification has been suggested, with a need for irrigation, improved varieties, and increased availability of fertilisers and pesticides all being highlighted [39]. Access to training, technology and finance is, as is so often the case, a major barrier. In India, the successful introduction of new heat-tolerant and disease-resistant potato varieties has been helped by direct participation of farmers in what is selected, and where and how it is used. With malnutrition a big and still-growing problem in India [41], the provision of biofortified potato plants (ones that produce crops especially enriched in key nutrients) also has huge potential—more than half of children in India are currently at risk of vitamin A deficiency and its resulting health problems, including childhood blindness [42].

The West Lothian chip buttie that we eat in years to come may be smaller and the prices higher but, for millions around the world, failure to realise a climate-smart future for the humble spud may mean there is no supper at all.

References

1. ScotGov. *Scottish Health Survey 2016.* Population Health Directorate, Scottish Government. https://www.gov.scot/publications/scottish-health-survey-2016-volume-1-main-report/ (2016).

2. Leith, W. The secret life of chips. *The Guardian.* https://www.theguardian. com/lifeandstyle/2002/feb/10/foodanddrink.features2 (2002).

3. FAO. *Potato.* http://www.fao.org/land-water/databases-and-software/crop-information/potato/en/ (2018).

4. Nunn, N. & Qian, N. The potato's contribution to population and urbanization: Evidence from a historical experiment. *Q. J. Econ.* **126**, 593–650 (2011).

5. Potatopro. *The Potato Sector.* https://www.potatopro.com/world/potato-statistics (2019).

6. Ritchie, H. Global potato production, 2014. *Ourworldindata.org.* https://ourworldindata.org/grapher/potato-production (2018).

7. AHDB. *AHDB Potatoes Variety Database.* https://potatoes.ahdb.org.uk/seed-exports/varieties (2019).

8. FAO. *FAO Success Stories on Climate-Smart Agriculture.* Food and Agriculture Organization of the United Nations. http://www.fao.org/3/a-i3817e.pdf (2014).

9. AHDB. *Common Scab.* https://potatoes.ahdb.org.uk/media-gallery/detail/13214/2638 (2019).

10. Obidiegwu, J. E., Bryan, G. J., Jones, H. G. & Prashar, A. Coping with drought: Stress and adaptive responses in potato and perspectives for improvement. *Front. Plant Sci.* **6**, 542 (2015).

11. Hodge, C. *Minimising Damage.* Potato Council. https://potatoes.ahdb.org. uk/sites/default/files/publication_upload/Damage Brochure 131121 Low Resolution.pdf (2014).

12. Williams, A. G., Audsley, E. & Sandars, D. L. Environmental burdens of producing bread wheat, oilseed rape and potatoes in England and Wales using simulation and system modelling. *Int. J. Life Cycle Assess.* **15**, 855–868 (2010).

13. Ponsioen, T. & Blonk, H. *Case Studies for More Insight into the Methodology and Composition of Carbon Footprints of Table Potatoes and Chips* (Blonk Environmental Consultants: Gouda, The Netherlands, 2011).

14. chippychat.co.uk. *Fish & Chip Industry Announce Standard Portion Size Recommendations.* https://www.chippychat.co.uk/fish-chip-industry-announce-standard-portion-size-recommendations/ (2018).

15. Schmidt, J. H. Life cycle assessment of five vegetable oils. *J. Clean. Prod.* **87**, 130–138 (2015).

16. Andersson, K., Ohlsson, T. & Olsson, P. Screening life cycle assessment (LCA) of tomato ketchup: A case study. *J. Clean. Prod.* **6**, 277–288 (1998).

17. WRAP. Household food and drink waste in the United Kingdom 2012. *Waste and Resource Action Programme.* http://www.wrap.org.uk/sites/files/wrap/hhfdw-2012-main.pdf.pdf (2013).

18. Smithers, R. Nearly half of all fresh potatoes thrown away daily by UK households. *The Guardian.* https://www.theguardian.com/environment/2017/nov/08/nearly-half-of-all-fresh-potatoes-thrown-away-daily-by-uk-households (2017).

19. BBC. *Leftover Potato Recipes.* https://www.bbcgoodfood.com/recipes/collection/leftover-potato (2018).
20. MetOffice. *UK Weather Summary, March 2012.* https://www.metoffice.gov.uk/climate/uk/summaries/2012/march (2012).
21. AHDB. *Blackleg and Soft Rots.* https://potatoes.ahdb.org.uk/agronomy/plant-health-weed-pest-disease-management/blackleg-and-soft-rot (2018).
22. Donnelly, J. The Irish Famine. *BBC Online.* http://www.bbc.co.uk/history/british/victorians/famine_01.shtml (2017).
23. Forestphytophthoras.org. *Phytophthora Basics.* http://forestphytophthoras.org/phytophthora-basics (2019).
24. CABI. *Phytophthora infestans (Phytophthora blight).* CABI Invasive Species Compendium. https://www.cabi.org/isc/datasheet/40970 (2018).
25. Singh, B. P., Dua, V. K., Govindakrishnan, P. M. & Sharma, S. *Climate-Resilient Horticulture: Adaptation and Mitigation Strategies* 125–135 (Springer, 2013).
26. MetOffice. *Land Projections Maps: Probabilistic Projections.* https://www.metoffice.gov.uk/research/collaboration/ukcp/land-projection-maps (2018).
27. MetOffice. *UK Weather Summary, July 2018.* https://www.metoffice.gov.uk/climate/uk/summaries/2018/july (2018).
28. Nicolson, N. Potato fields under pressure as drought conditions continue. *The Courier.* https://www.thecourier.co.uk/fp/business/farming/farming-news/686157/potato-fields-under-pressure-as-drought-conditions-continue/ (2018).
29. Tholhuijsen, L., Allison, R. & Wills, R. How the drought has affected potato yields and quality. *Farmers Weekly.* https://www.fwi.co.uk/arable/harvest/how-the-drought-has-affected-uk-and-european-spud-yields (2018).
30. Kindred, A. Chips are down. *The Sun.* https://www.thesun.co.uk/news/7292343/potato-crisis-chips-european-drought/ (2018).
31. Knox, J., Daccache, A., Weatherhead, K. & Stalham, M. *Climate Change and Potatoes: The Risks, Impacts and Opportunities for UK Potato Production.* Potato Council (AHDB) and Cranfield University. http://www.potato.org.uk/sites/default/files/publication_upload/CC impacts potatoes_Final_20Sept2011.pdf (2011).
32. Knox, J. W., Daccache, A., Weatherhead, E. K. & Stalham, M. Climate change impacts on the UK potato industry. *Potato Council (AHDB) Report No. 2011/3.* https://potatoes.ahdb.org.uk/sites/default/files/publication_upload/20113 Climate Change R404 Final R404.pdf (2011).
33. Daccache, A. *et al.* Climate change and land suitability for potato production in England and Wales: Impacts and adaptation. *J. Agri. Sci.* **150**, 161–177 (2012).
34. Raymundo, R. *et al.* Climate change impact on global potato production. *Eur. J. Agron.* **100**, 87–98 (2018).

35. Steele, C. *Irrigation and Water Use (Best Practice Guide for Potatoes)*. Potato Council (AHDB). https://potatoes.ahdb.org.uk/sites/default/files/publication_upload/Irrigation for potatoes_0.pdf (2013).

36. Jones, D. Potato yields and soils improved by green waste compost on Norfolk farm. *Farmers Weekly*. https://www.fwi.co.uk/arable/crop-management/nutrition-and-fertiliser/potato-yields-soils-improved-compost (2018).

37. George, T. S., Taylor, M. A., Dodd, I. C. & White, P. J. Climate change and consequences for potato production: A review of tolerance to emerging abiotic stress. *Potato Res.*, 1–30 (2018).

38. Kanter, J. Potato production and climate change. *spudsmart.com*. https://spudsmart.com/potato-production-climate-change/ (2018).

39. Hengsdijk, H. & Verhagen, A. *Linking Climate Smart Agriculture and Good Agriculture Practices: Case Studies on Consumption Potatoes in South Africa, the Netherlands and Ethiopia*. (Plant Research International, 2013).

40. Muchaba, T. *Are These the Climate-Smart Potatoes?* CGIAR Climate Change, Agriculture & Food Security. https://ccafs.cgiar.org/news/are-these-climate-smart-potatoes - .XBfbly2cbUp (2017).

41. Ritchie, H., Reay, D. S. & Higgins, P. Quantifying, projecting and addressing India's hidden hunger. *Front. Sustain. Food Sys.* **2**, 11 (2018).

42. cipotato.org. *Growing Collaboration for the Benefit of India's Farmers and Families*. https://cipotato.org/blog/collaboration-benefit-indias-farmers-families/ (2018).

CHAPTER 13

Climate-Smart Cod

Abstract We eat over 150 million fish & chip meals each year in the UK. For fish and shellfish more generally, average consumption is now around 20 kilograms per person each year in the UK and US, but this is dwarfed by consumption in Asia—in China, it is nudging 35 kilograms per person a year for a population of over 1.3 billion. Cod remains the common source of British fish and chips, but stocks have experienced intense overfishing in the past and many have still not recovered. Warming in the North Sea has pushed Atlantic cod stocks further north, and most cod eaten in the UK now comes from Iceland and Norway. An uncooked cod fillet has a carbon footprint of around 300 grams. For a full portion of deep fried fish and chips, this rises to a kilogram of emissions. Cutting household wastage and improving the efficiency of cooking can reduce this. At sea, a switch from bottom trawling to pelagic fishing would reduce fuel use and also help to protect cod stocks and their food supplies on the ocean bed. For all fisheries, increased protection from overexploitation and pollution can lend them greater resilience in the face of climate change.

Keywords Fish & chips • Shellfish • Bottom trawling • Sea ice • Grand Banks • North Sea • Iceland • Carbon footprint • Aquaculture

Chips are wonderful. A freshly battered piece of fish to go with them is nirvana. Fish & chips remains a takeaway staple across the land, and each

© The Author(s) 2019
D. Reay, *Climate-Smart Food*,
https://doi.org/10.1007/978-3-030-18206-9_13

year, we wolf down over 150 million of these deep-fried delights [1]. Globally our appetite for fish and seafood continues to grow apace. In the US, Canada and the UK, there has been a steady increase in consumption since World War Two, and today, the annual average is about 20 kilograms of fish and shellfish per person. Across large parts of Asia, however, demand has exploded. In China, per capita consumption has grown more than eightfold since the 1960s and is now nudging 35 kilograms a year (Fig. 13.1). With a population of over 1.3 billion, that's an awful lot of fish.

Despite ballooning demand, the amounts of wild fish caught each year appear to be levelling off (at around 90 million tonnes a year). Filling the yawning supply gap has come aquaculture. From salmon farms to shrimp ponds, oyster rafts to carp pools, the marine and freshwater farming of fish and shellfish has become big business around the world. In 1960 it produced just 2 million tonnes of food globally. By 2015 it was topping 100 million tonnes a year and had comfortably overtaken wild capture as the main source of all things fishy in the world's food baskets [3].

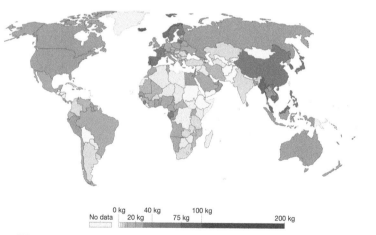

Fish and seafood consumption per capita, 2013

Annual consumption of fish and seafood per person per year, measured in kilograms. Data is inclusive of all fish species and major seafood commodities, including crustaceans, cephalopods and other mollusc species. Data is based on per capita food supply at the consumer level, but does not account for food waste at the consumer level.

Source: UN Food and Agriculture Organization (FAO)   OurWorldInData.org/meat-and-seafood-production-consumption/ • CC BY

Fig. 13.1 Global fish and shellfish consumption per capita in 2013 (Source: Hannah Ritchie, Our World in Data) [2]. Available at: https://ourworldindata. org/grapher/fish-and-seafood-consumption-per-capita

Here in Britain most fish & chip suppers are still courtesy of wild catches of the wide-mouthed big-bellied fish that is the cod [4]. Cod has been a sought-after dinner for centuries and part of European diets since the stone age [5]. There are three main commercial species: the Atlantic cod (the source of our supper), the Pacific cod and the Greenland cod. They are slow swimmers, and their capacious maws mean they can hoover up everything from shrimp and other fish, to worms, shellfish and even sea urchins [6].

Atlantic cod inhabit surface waters right down to depths of more than 200 metres (sometimes as deep as 600 metres) and prefer water temperatures between 2 and 8 degrees Celsius. Females reach maturity at three to six years, and each produce several million eggs—the record from a single female is nine million—which are released over the ocean floor. The eggs hatch within about a month and, after a few more months as drifting larvae, the fry then settle down for a life mostly focussed on food gathering near the seabed. They can live for over 25 years [7]. Fish over 6 feet in length and weighing in at more than 200 pounds have been reported in the past, but such leviathans are likely long gone, and the average size of cod today is much diminished from its pre-twentieth-century heyday.

Like 90 per cent of the cod eaten in Britain, our own crisply battered cod fillets come from fish caught in the icy waters around Iceland and in the Barents Sea. There are still some cod closer to home, such as in the colder corners of the North Sea and off the northwest coast of Scotland [8], but the bulk of the supply for our 10,000 or so fish & chip shops is now imported from Iceland and Norway [9].

To catch their quarry, trawlers use a range of techniques, including long lines that carry thousands of baited hooks attached at intervals along a buoyed main line up to 30 miles in length, and nets that either target fish in the middle or upper waters (pelagic) or those near the bottom (demersal) [10]. A traditional one for cod is the otter trawl, where cone-shaped nets are drawn across the sea bed with the net mouth kept open using rectangular steel boards—the otter boards—attached on either side [11]. Any cod in its path are directed into its funnel-like body and end up held in the aptly named 'cod end' at the back. Once hauled on board, the fish are usually gutted and covered with ice ready for the journey to shore. Back in ports like Reykjavík they are filleted and boxed up for refrigerated transport by sea to the deep-fat fryers of Britain.

Wild caught cod effectively have a carbon footprint of zero until they are caught, but after that the emissions can rapidly mount up. The catching phase is the big one. Fuel (usually diesel) is burned getting to and from the fishing grounds, as well as during trawling itself. Bottom trawls are especially fuel-intensive as the heavy net and otter boards need a lot of power to drag them along. The trawls can also catch many non-target species and have a devastating impact on the seabed ecosystems they plough through.

More energy use, and so emissions, arises from ice making, refrigeration and running the boats. For every tonne of edible cod, this catching phase results in the equivalent of over a tonne of carbon dioxide [12, 13]. Along with processing, onwards transport to the UK and still more refrigeration, each cod fillet arrives at the chip shop with a carbon footprint of around 300 grams [12]. As with chips, this is bumped up by the oil and energy then used for frying. The indispensable batter brings the life-cycle footprint of one crispy, deep-fried fillet to half a kilogram.

Our completed paper-wrapped bundle of battered fish fillet and generous portion of chips therefore tips the climate scales at about 1 kilogram. Rather weighty, but still a slim line climate supper compared to many of the plastic-encased meals on offer from our supermarkets—a ready-meal dinner of beef burritos, for example, has a carbon hoofprint five times that of our fish supper [14].

Given its takeaway nature, the options at home for us to directly cut the life-cycle emissions involved in fish & chips appear limited. The packaging is rarely recyclable due to contamination by oil, and other than warming up plates in the anticipation of the meal's arrival, there isn't much home energy used in its consumption. As ever, where we can do most is in avoiding waste.

In the UK we throw away an estimated 35,000 tonnes of fish and shellfish each year. This includes guts, bones and shells, but over 80 per cent is deemed avoidable waste, due mainly to the food not being used in time or because too much is cooked or served [15]. It includes everything from tinned sardines and salmon steaks, to fish fingers and our battered cod fillets. Some products will have much lower carbon footprints than cod—farmed mussels in Scotland are responsible for under 300 grams of emissions for each kilogram of shiny-shelled treats [16]. Others will be much more carbon-intensive. Spanish tuna, for instance, has a climate footprint of around 2 tonnes for every tonne of fish that gets to market [17]. Overall, the annual British wastage of these delicious fruits of the sea equates to around 30,000 tonnes of avoidable emissions.

*   *   *

Over the centuries many fish species have had their share of trials and tribulations at the hands of humankind, but few can rival that of the Atlantic cod and its near-annihilation on Canada's Grand Banks. The Grand Banks are a group of shallow underwater shelves off the coast of Newfoundland that sit at the oceanic meeting point of the cold Labrador current and the warm Gulf Stream. The abundance of nutrients provided by these mixed-up waters makes for a super-productive sea and so a smorgasbord of crustaceans, worms and small fish. For centuries cod have been filling their boots.

In the 1600s fishermen reported that the cod shoals were so dense that they had trouble rowing through them [18]. Newfoundland's fishing industry grew fast as increasing numbers of boats came from Europe and Scandinavia to reap the seemingly endless catch. By the 1950s hulking factory trawlers, each able to catch and freeze many hundreds of tonnes of fish, were ploughing the rich waters for weeks at a time. Their frozen harvest was sailed back to ports all over the world, before the boats returned for another helping. The Grand Banks' cod stocks were massive, but even they could not withstand this industrial extermination. Each new year brought a new record catch, with 1968 seeing over 800,000 tonnes of fish taken. Those heights were never reached again. Despite bigger boats and nets, the catch began to slide. By the early 1970s it was down to 300,000 tonnes a year. The Canadian Government intervened. Not to protect the cod, but to extend the exclusive area over which it controlled fishing and so help its own fleet get in on the cod bounty.

Yet more factory trawlers and processing factories were built and a new homegrown pressure on stocks was ratcheted up. Scientific advice at the time was that a 250,000 tonne-a-year catch was sustainable. By the late 1980s, that advice was found badly wanting. Catches nose-dived, yet fishing continued until there was virtually nothing left to catch. By the 1990s estimated cod stocks were just 1 per cent of their 1960s levels, and only 2,000 tonnes of breeding-age fish remained. The industry collapsed. Some 30,000 fishermen found themselves drawing benefits instead of nets, and another 15,000 in jobs like ship building and fish processing ended up out of work [18].

The disaster for people and cod alike that was the Grand Banks at least served as a warning to governments of the folly of setting politically motivated quotas. It was also a brutal wake-up call for fisheries science. Since then, evidenced-based policy and improved fisheries management have helped to head-off more catastrophic declines. Though sometimes only just. In 2006 the North Sea cod populations were rescued from the brink of collapse by a raft of measures such as larger mesh sizes in nets (to let younger fish escape), boat decommissioning and bans on fishing in nursery areas [9]. Today's stocks have recovered to a 35-year high (albeit from a desperately low baseline) but, just as a sustainable future for Britain's North Sea cod fishery appears within our grasp, climate change is snatching it away.

Temperature affects all the life-cycle stages of fish, including those of cod [19]. As oceans warm, so their reproductive success, food supplies, growth rates and distribution can be altered [20]. The effects vary depending on where the cod populations are. In the 1920s and 1930s, for instance, North Atlantic cod off the shores of West Greenland—the coldest northern reaches of their normal range—appeared to benefit from a warming sea. Down at their warmer southern limits, past warming has often been less welcome and numbers have declined [21, 22]. One key issue seems to be changing food supply as the waters warm and in particular the types and numbers of copepods (small, free-swimming crustaceans) and other zooplankton that the cod rely on in their early years. As water temperatures rose in the second half of the twentieth century, the distribution of cold-water copepods favoured by young North Sea cod retreated northwards [23].

In the twenty-first century, the North Sea, like most ocean areas, is set to warm a whole lot more. By the 2090s the average sea surface temperature could rise by a further 3 degrees Celsius (it is already about 1 degree Celsius warmer than in the 1960s [24, 25]). This warming is projected to be most rapid in southern waters and during the summer months, when the seas become more stratified and a stable warm layer can form near the surface. Warming at depth will be slower, but the southern North Sea is shallow and its cod population faces the prospect of dwindling food supplies in waters that are becoming progressively warmer, less saline (due to ice melt and increased rainfall run-off), and more acidic (due to more carbon dioxide) [26].

Even without climate change, in the absence of sustainable fisheries management, North Sea cod stocks by the middle of the century are predicted to decline. With climate included, they would plummet. A moderate warming scenario suggests that the numbers of cod surviving long enough to join the adult population will drop by almost one-third. Under a rapid warming scenario, such cod recruitment in the North Sea could fall by over 95 per cent [27]—disaster on a grand (Banks) scale.

North Sea cod ranges are already retreating to colder and deeper waters [28]. The picture for hundreds of other cold-water fish species around the world is a similar one of pole-ward retreat in the face of overheating oceans [29]. For some, populations at the chillier extremities of their species' range may benefit—warming of 1 to 2 degrees Celsius is expected to expand numbers of Atlantic cod around Greenland and in the iciest reaches of the Barents Sea [19, 30], while southern waters are becoming home to a host of new invaders, including sardines and cuttlefish, nosing up from the south [31].

For species like polar cod—a boreal cousin of the Atlantic cod but with a much lower and tighter preferred temperature range of −1 to 2 degrees Celsius—warming waters may deliver an extra whammy in the form of retreating sea ice and loss of the spring and summer nursery this ice provides for their larvae and fry [20].

* * *

Making plants and animals healthier in the first place is a fundamental step in building climate resilience. The same is true of our oceans. The ability of global fisheries and wider marine ecosystems to cope with the challenges of warming and acidification, of hurricanes and sea level rise, can be greatly increased by ensuring they are in the fittest state possible to do so. The future of coral reef systems, for instance, ultimately rests on whether we can keep warming to below 1.5 degrees Celsius by the end of the century, but in the meantime avoiding damage from issues like coastal run-off and eutrophication can at least lighten their load. For commercial fish species like cod, ensuring healthy populations through sustainable fisheries management—protecting breeding grounds and enforcing strict quotas—is similarly vital if they are to weather the coming storm [30].

There's a risk that emissions from the cod trawler fleet will increase in the future as boats travel further and further from port to find fish [32]. Again, this risk could be reduced by boosting fisheries protection and so maintaining stocks closer to home. The boats themselves can cut emissions through improving engine efficiencies, using lower carbon fuels, and adopting alternative fishing methods. Switching away from the destructive and energy-intensive practice of bottom trawling is a prime example of this. By using different fishing methods, like drift netting and pelagic (upper water) trawls, the amount of fuel used per kilogram of fish caught can be cut by 80 per cent [32, 33]. It also means less damage to seabed communities. As these provide much of the food cod and other commercial fish species rely on, their protection offers a longer-term triple-win of greater productivity of the fishery, lower emissions from the boats and enhanced climate change resilience of the wider ecosystem. The nets may have fewer cod when they are pulled on board, but in reality, the most climate-smart option for our future fish suppers may well be to embrace other fish species anyway.

Increasing numbers of restaurants and takeaways now offer alternatives like pollack, coley or hake as their battered dish of the day [34]. Warmer-water invaders like the red mullet and John Dory offer the prospect of new markets for the fishing fleets and new culinary delights for the dinner table [31]. Cod farming using seawater ponds or floating coastal pens may also help take some of the strain off wild populations. Though still a minor part of the global market, it is growing fast, with output in Norway having already increased to over 16,000 tonnes in 2008 [35]. In theory such managed production can avoid the large 'catch phase' emissions of wild stocks and give more resilience to climate change. Larvae and young fish can be reared under controlled conditions and the fish can be supplied with all the feed they need to grow fast [36].

With rearing ponds and cages being close to shore, however, warming risks may be even greater and concerns have been raised about farmed cod escaping (they are inquisitive fish and good at finding their way out of nets). Where these escapees interact with wild stocks, there is a risk of disease spread and of interbreeding that reduces the wild cod gene pool [35]. The low-carbon credentials of such cod aquaculture are also questionable as the emissions saved on the high seas are partially or wholly offset by those of producing the fish feed and of powering water pumps, filters and the rest.

Ultimately the climate-smart fillet in our fish supper will be one sourced from a system, whether wild or farmed, that takes full account of the impacts of a changing climate. It will also be one that strives to limit its own role in accelerating those impacts. Wild fisheries and aquaculture directly employ over 50 million people around the world, with a tenth of the global population deriving their livelihoods in one way or another from fish and shellfish. Nine out of ten people working in capture fisheries are in small-scale artisanal operations. Providing them with good support and advice, alongside sustainable finance and evidence-based regulation, can help ensure that our growing reliance on fish and shellfish for global food security is climate-proofed [37].

REFERENCES

1. Federationoffishfriers.co.uk. *Fish and Chips, Facts and Figures.* http://www. federationoffishfriers.co.uk/pages/facts-and-figures-603.htm (2019).
2. Ritchie, H. Global fish and shellfish consumption per capita, 2013. *Ourworldindata.org.* https://ourworldindata.org/grapher/fish-and-seafood-consumption-per-capita (2018).
3. Ritchie, H. Global production from capture fisheries versus from aquaculture, 1960–2015. *Ourworldindata.org.* https://ourworldindata.org/grapher/capture-fisheries-vs-aquaculture-farmed-fish-production (2018).
4. SeaFish. *The Fish We Eat With Our Chips—The Facts.* https://www.seafish. org/media/390797/the fish we eat with our chips - the facts - master jan 2011.pdf (2018).
5. msc.org. *What is Cod?* https://www.msc.org/what-you-can-do/eat-sustainable-seafood/fish-to-eat/cod (2019).
6. Kennedy, J. Atlantic Cod (*Gadus morhua*). *thoughtco.com.* https://www.thoughtco.com/atlantic-cod-gadus-morhua-2291590 (2017).
7. UCD.ie. *Cod Biology.* http://www.ucd.ie/codtrace/codbio.htm (2002).
8. seafoodscotland.org. *Responsible Sourcing—Cod.* http://www.seafoodscotland.org/ja/responsible-sourcing/top-species/cod.html - biology (2019).
9. Carrington, D. Sustainable British Cod on the Menu After Stocks Recover. *The Guardian.* https://www.theguardian.com/environment/2017/jul/19/sustainable-british-cod-on-the-menu-after-stocks-recover (2017).
10. britishseafishing.co.uk. *Commercial Fishing Methods.* http://britishseafishing.co.uk/commercial-fishing-methods/ (2019).
11. mcsuk.org. *How Fish are Caught.* https://www.mcsuk.org/media/seafood/Fishing_Methods.pdf (2019).
12. Smárason, B., Viðarsson, J., Thordarson, G. & Magnúsdóttir, L. *Life Cycle Assessment of Fresh Icelandic Cod loins* (Reykjavík: Matís, 2014).
13. Sund, V. *Environmental Assessment of Northeast Arctic Cod Caught by Long-Lines and Alaska Pollock Caught by Pelagic Trawls* (SIK Institutet för livsmedel och bioteknik, 2009).
14. Tesco. *Product Carbon Footprint Summary.* https://www.tescoplc.com/assets/files/cms/Tesco_Product_Carbon_Footprints_Summary(1).pdf (2012).
15. WRAP. Household food and drink waste in the United Kingdom 2012. *Waste and Resource Action Programme.* http://www.wrap.org.uk/sites/files/wrap/hhfdw-2012-main.pdf.pdf (2013).
16. Fry, J. M. *Carbon Footprint of Scottish Suspended Mussels and Intertidal Oysters.* Scottish Aquaculture Research Forum (SARF). http://www.sarf.org.uk/cms-assets/documents/43896-326804.sarf078 (2012).

17. Hospido, A. & Tyedmers, P. Life cycle environmental impacts of Spanish tuna fisheries. *Fish. Res.* **76**, 174–186 (2005).
18. britishseafishing.co.uk. *The Collapse of the Grand Banks Cod Fishery.* http://britishseafishing.co.uk/the-collapse-of-the-grand-banks-cod-fishery/ (2019).
19. Drinkwater, K. F. The response of Atlantic cod (*Gadus morhua*) to future climate change. *ICES J. Mar. Sci.* **62**, 1327–1337 (2005).
20. Dahlke, F. T. *et al.* Northern cod species face spawning habitat losses if global warming exceeds 1.5° C. *Sci. Adv.* **4**, eaas8821 (2018).
21. Mieszkowska, N., Sims, D. & Hawkins, S. *Fishing, Climate Change and North-East Atlantic Cod Stocks* (2007).
22. O'brien, C. M., Fox, C. J., Planque, B. & Casey, J. Fisheries: Climate variability and North Sea cod. *Nature* **404**, 142 (2000).
23. Beaugrand, G., Brander, K. M., Lindley, J. A., Souissi, S. & Reid, P. C. Plankton effect on cod recruitment in the North Sea. *Nature* **426**, 661 (2003).
24. Wiltshire, K. H. & Manly, B. F. The warming trend at Helgoland Roads, North Sea: Phytoplankton response. *Helgol. Mar. Res.* **58**, 269 (2004).
25. Carbonbrief. Mapped: How every part of the world has warmed—And could continue to warm. *Carbonbrief.org.* https://www.carbonbrief.org/mapped-how-every-part-of-the-world-has-warmed-and-could-continue-to-warm (2018).
26. Schrum, C. *et al. North Sea Region Climate Change Assessment* 175–217 (Springer, 2016).
27. Clark, R. A., Fox, C. J., Viner, D. & Livermore, M. North Sea cod and climate change–modelling the effects of temperature on population dynamics. *Glob. Chang. Biol.* **9**, 1669–1680 (2003).
28. Engelhard, G. H., Righton, D. A. & Pinnegar, J. K. Climate change and fishing: A century of shifting distribution in North Sea cod. *Glob. Chang. Biol.* **20**, 2473–2483 (2014).
29. Morley, J. W. *et al.* Projecting shifts in thermal habitat for 686 species on the North American continental shelf. *PLoS One* **13**, e0196127 (2018).
30. Kjesbu, O. S. *et al.* Synergies between climate and management for Atlantic cod fisheries at high latitudes. *Proc. Natl. Acad. Sci.* **111**, 3478–3483 (2014).
31. McKie, R. Cod and haddock go north due to warming UK seas, as foreign fish arrive. *The Observer.* https://www.theguardian.com/environment/2017/sep/02/fish-conservation-foreign-species-uk-waters-climate-change (2017).
32. EC. The carbon footprint of fisheries. *Seas at Risk.* European Commission. https://energyefficiency-fisheries.jrc.ec.europa.eu/c/document_library/get_file?uuid=924c1ba8-94af-440d-94cb-f9cb124d2d57&groupId=12762 (2009).
33. Ziegler, F. & Hansson, P.-A. Emissions from fuel combustion in Swedish cod fishery. *J. Clean. Prod.* **11**, 303–314 (2003).

34. McDonald, K. Sustainable seafood: How to eat fish without destroying the ocean. *iNews.* https://inews.co.uk/news/uk/sustainable-seafood-eat-fish-destroying-ocean-cod-king-prawn-replacement/ (2018).
35. Fouché, G. Rapid growth in European cod farming prompts fears from green groups. *The Guardian.* https://www.theguardian.com/environment/2009/jun/22/european-cod-farming-norway (2009).
36. thefishsite.com. *How to Farm Atlantic Cod.* https://thefishsite.com/articles/cultured-aquatic-species-atlantic-cod (2012).
37. FAO. *Climate-Smart Fisheries and Aquaculture.* Food and Agriculture Organization of the United Nations. http://www.fao.org/climate-smart-agriculture-sourcebook/production-resources/module-b4-fisheries/b4-overview/en/ (2017).

CHAPTER 14

Climate-Smart Champagne

Abstract There are three billion bottles of fizz drunk each year, and 80 per cent of these come from Europe. Champagne has a carbon footprint of around 2 kilograms of emissions per bottle. The UK alone wastes 40,000 tonnes of wine each year, equivalent to almost 100,000 tonnes of greenhouse gas emissions from drink that ends up down the drain. The Champagne region has already seen a more than 1-degree-Celsius rise in average temperatures and may see over 5 degrees Celsius of warming by the end of the century. Wine growers can adapt to a changing climate through irrigation, pruning techniques and the use of alternative grape varieties, but many, including Champagne, rely on specific vines and conditions. In the longer term, some growers will have to either move their vineyards to cooler locations or give up on the wines they have produced for generations. Either way, the places our sparkling wine is produced and what it is called are set to change radically in the coming decades.

Keywords Sparkling wine • Cavea • Cremant • Sekt • Prosecco • Botrytis • Terroir • Appellation • Pierce's disease • Viticulture • Carbon footprint

The crowning glory of our day's food is that most giggly of pleasures: champagne. As an accompaniment to a Friday night fish supper, there is

© The Author(s) 2019
D. Reay, *Climate-Smart Food*,
https://doi.org/10.1007/978-3-030-18206-9_14

nothing finer. Though its cheaper cousins Prosecco and Cava are our own more usual celebration tipple, for a big event like a birthday, an anniversary or simply (to be honest) the end of a bad week, the cork-popping joys of champagne are hard to beat.

Once the preserve of the rich and privileged, sparkling wines like champagne have now become an attainable luxury loved by millions. Large-scale production and lower prices have seen sparkling wine consumption rocket—in the UK we now toast our way through over 100 million bottles each year. Around one-third of these are true French champagne, with the rest being Prosecco, Cava and the like. Rather excitingly for English growers, the fourth biggest-selling sparkling wine in the UK is now homegrown [1].

Europe has always been at the epicentre of global fizz production and still produces 80 per cent of the three billion bottles popped globally. Italy and its ubiquitous Prosecco lead the way, with French champagne and Cremant, German Sekt and Spanish Cava being the other bubbly big league players [2]. Since 1990 global sales have surged, leaving stocks of more traditional wines gathering dust—every tenth bottle of wine we now buy is fizz. A good part of this surge is down to drinkers in North America and the UK. France may still sit unsteadily atop the world wine-drinking league (Fig. 14.1), but Britain and the US are the world's biggest fizz importers and the past quarter of century has seen their thirst more than triple [2].

Sparkling wine, like all wine, is a well-established canary in the climate change coalmine. The global nature of its production, the specific climate needs of the various vines, and the often centuries-long records of cultivation and harvest, give a rich view of how climate has changed in the past and what may be in store for the future [4]. To explore the climate risks and responses for our beloved fizz, we are really going to push the boat out. Tonight it must be real French champagne, and it must be one of the very best. It must be Bollinger.

> *I drink Champagne when I'm happy and when I'm sad. Sometimes I drink it when I'm alone. When I have company I consider it obligatory. I trifle with it if I'm not hungry and drink it when I am. Otherwise, I never touch it—unless I'm thirsty.* Lily Bollinger, House of Bollinger Champagne [5]

Bollinger is one of the great French champagne houses, and like Moet & Chandon, Veuve Clicquot, Krug and Taittinger, they are strictly

Wine consumption per person (litres pure alcohol), 2014
Average per capita consumption of wine, as measured in litres of pure alcohol per year.

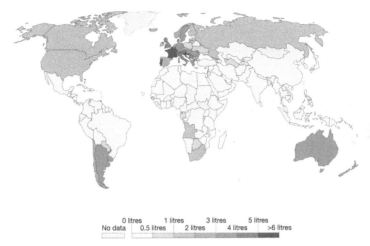

	0 litres	1 litres	3 litres	5 litres	
No data	0.5 litres	2 litres	4 litres	>6 litres

Source: WHO Global Health Observatory (GHO)

Fig. 14.1 Global wine consumption per capita (litres of pure alcohol) in 2014 (Source: Hannah Ritchie, Our World in Data) [3]. Available at: https://ourworldindata.org/grapher/wine-consumption-per-person

confined to producing their wine from vines grown on the undulating slopes of north eastern France's Champagne region. Well-drained chalky soils, together with a cool climate, moderate rainfall and the long-established wine growing practices and culture of the region make for the Champagne terroir—the distinctive physical, biological and cultural attributes of a wine growing area [6, 7].

To qualify as true champagne, the vines must be grown in this specific area and to a strict set of rules (the coveted appellation). The appellation includes which varieties can be used—Chardonnay and Pinot Noir are usual, how much can be grown, pruning techniques, grape alcohol contents and fermentation methods. Achieving the champagne designation means navigating a maze of regulations, but the rewards can be astounding. With such huge global demand and only a limited area to produce it, the very best bottles may fetch in excess of $1,000. An 1820 Juglar Cuvee will set you back over $40,000 (complete with barnacle encrustations from its time aboard a sunken ship) and the current record is $2 million for a bottle called Taste of Diamonds, though its sky-high price is probably more to do with its handcrafted gold label featuring a 19-carat diamond [8].

Unsurprisingly land prices in Champagne are stratospheric too, averaging around half a million dollars per hectare [9]. Our own pricey, though diamond-free, bottle of Bollinger therefore began life as a mixture of grapes grown in this super-select corner of global viticulture.

*   *   *

Grape vines like it sunny. In general they need long growing seasons (150–180 days) relatively low rainfall and humidity, but still with enough soil moisture to keep them happy through the summer [10]—a tough balancing act for growers as temperatures rise and rainfall patterns change. Vines remain dormant at below 10 degrees Celsius, and most varieties tend to do best in areas where the average growing season temperatures are somewhere between 13 and 21 degrees Celsius—it's nigh on impossible to produce good wines in the tropics and sub-tropics [4, 11]. The Champagne region is near the northern limits for reliable viticulture, and it is the slow-growing nature of the grapes there that helps to produce the crisp acidic characteristics needed for making champagne [7].

Vines take two or three years from planting before they start producing grapes and, as they are climbing plants, they need support as they grow [12]. The succulent bags of sugar that are grapes inevitably attract a lot of pests and diseases. Pinot noir—a staple grape for champagne—is prone to fungal attack by powdery and downy mildews, the common grey mould of botrytis, and an array of viruses and sap-feeding pests. One of the most worrying is Pierce's disease, caused by a bacteria that invades the vines when insects feed on them. Leaves of infected plants first begin to turn red or yellow, their grapes shrivel and a growing carpet of dead leaves accumulates below the under-siege plants. There is no cure [13]. Frost damage is also a perennial risk at high latitudes and altitudes—a hard spring frost when the new buds are just forming can wipe out that year's grape harvest [14].

After about eight years of care and attention, the vines should hit full grape production. Regular winter pruning, weeding, fertiliser application and pest control are usually needed throughout. With luck the vine can then go on to produce good grapes for many decades—they can live for upwards of 70 years, and some ancient vineyards in Slovenia still produce grapes from vines planted four centuries ago [12].

Correctly judging when to harvest is critical to the success of the wine produced. Growers are careful to wait until the sugar content and acidity levels are just right—in a very warm year, the perfect time may come much earlier, in a cold year, much later [15].

Once harvested, either by hand or by machine, the grapes are immediately taken for initial processing. Stems are removed, the grapes crushed, and the resulting juice (called must) transferred to fermenters where added yeast gets to work converting the sugars to alcohol and producing streams of carbon dioxide bubbles in the process. For red wine the grape skins are fermented too, for white they are removed. Once fermentation is complete, the raw young wine is clarified—yeast and other particles are removed through filtering or settling them out—and racked into bottles and barrels for ageing and eventual sale. Champagne and other sparkling wines, however, have a crucial extra element. It's a yeasty trick once regarded by wine makers as annoying. Today it's the magic that makes our wine sparkle.

As grape juice ferments, the sugars are used up, and eventually, the yeast will run out of fuel. Sometimes though the new wine is racked too soon. If there is still enough sugar available, then the yeast will go on working and produce more alcohol, and lots more carbon dioxide. In sealed bottles the effect can be explosive, and even where the wine maker doesn't find their precious charge splattered across the cellar walls, the build-up of carbon dioxide will have turned the wine fizzy.

The inspiration to deliberately use this secondary fermentation to make champagne is often credited to the French Benedictine monk Dom Perignon, but in fact he spent years trying to work out how to avoid it. Instead, it was an English scientist called Christopher Merret who, in a paper to the Royal Society in 1662, outlined how 'sugar and molasses' could be added to new wine to make it sparkle.

By the middle of the eighteenth century, wine makers in Champagne were turning to exclusive production of sparkling wine using this sugar-adding technique, and in 1829, the House of Bollinger was founded. The champagne appellation demands secondary fermentation is done in the same bottles that we then buy, while for Prosecco and other sparkling wines, this step is more often done in large vats (making it less labour-intensive and so cheaper). Champagne bottles still explode sometimes, but stronger glass and more precise additions of sugar make this much rarer than in the early days—up until the 1830s, cellar workers routinely wore iron masks to protect them from random eruptions of glass, corks and bubbly [16].

Barring explosive failures, our own bottle of Bollinger will eventually make its way across the channel and find its way to our fridge. Each precious bottle has a life cycle carbon footprint of around 2 kilograms [17], with the bulk of this arising from grape growing (vineyard fuel, fertiliser and pesticide use) and the rest from the energy used in processing, packaging, transport and refrigeration. New world sparkling wines like those from Australia notch up further emissions—an extra 300 grams or so—due to the long-distance shipping required [18]. The three billion bottles of fizz consumed worldwide therefore have a carbon footprint in the region of six million tonnes a year. Just how much of all this we waste is unknown (not a drop in our house for sure).

Champagne's high price likely means less of it goes down the drain than most alcoholic drinks, but any wedding caterer or party host can testify to the fact that a lot still ends up decorating dance floors and carpets. For wine more broadly, the numbers on waste are instantly sobering. In the UK we throw away over 40,000 tonnes of wine each year at an estimated financial cost of £270 million. Even assuming sparkling wines suffer just half the wastage rates of other wines, this would still mean around 2,000 tonnes of dumped bubbly in the UK and the equivalent of some 4,000 tonnes of carbon dioxide emissions.

Almost all of such wine waste is deemed avoidable, the leading causes being the familiar ones of it getting old (not all wines age well) and too much being served. The rest of the wastage comes from personal preference and accidents—presumably, this last one increases in direct proportion to how much we've drunk [19]. Also familiar are ways to reduce this waste, including not over-buying and serving, and keeping an eye on drink-by dates. The big carbon savings for champagne, however, can be found further down the supply chain at the winery and vineyards. There too can be found the portents of a future climate that will redraw the global wine map and threaten even the hallowed diktats of the champagne appellation.

With Europe being the global powerhouse of sparkling wine production, it is severe weather and climate change impacts here that most threaten supplies worldwide. The intense heat wave of 2003 gave a fiery taste of the risks all farmers will face in the coming decades. In June of that year temperatures began to push past their normal levels across an expanding area of the continent. From Spain in the west to the Czech Republic in the east, and from northern Germany down to southern Italy, temperature records toppled as the heat intensified through July and into August. The all-time record in the UK fell on the 10th of August (hitting 38.1 degrees Celsius), and in France, temperatures surged past the 40 degrees Celsius mark and stayed there for weeks [20]. Along with an estimated 30,000 human casualties came big losses for many wheat, maize and livestock farmers [21]. In Champagne, the scorching weather meant a much earlier grape harvest amid concerns over heat stress to the vines [22], but the resulting vintage turned out to be a cracker. More recent heat waves, like that in 2018, again meant early harvest dates and making dawn raids on the vineyards before the heat became dangerous for pickers. Whether the resulting champagne is another good heat wave vintage won't be known for a while. What we do know is that the frequency and intensity of such extreme weather events is set to increase and that the cool climate envelope for growing champagne grapes is on the move.

As western Europe has warmed over the past 40 years, grape harvests across France have been occurring around 10 days earlier than the average for the preceding four centuries—harvests in the summer of 2003 were almost a month early [23]. By the middle of this century, severe heat waves like that of 2003 could be hitting us every other year [24]. The warming trend is tending to increase sugar levels in the grapes, making for wines that are sweeter and have a higher alcohol content [4]. As harvesting gets ever earlier, gaps may open up between the ideal harvest moment when the grapes have the right balance of sugar and acidity and the flavour moment when they will provide the specific taste required of fine wines [25]. Major champagne vines like Pinot Noir—the polar bear of wine in a changing climate [26]—are especially vulnerable as they like it cool and have a tight optimal temperature range.

Changing rainfall patterns and increasing temperatures will boost some pests and diseases too [27]. The small insects that transmit Pierce's disease are expected to expand northwards [28], and there are already concerns that vine-killing Black rot fungus is invading from the south as Europe warms [29].

By the middle of this century, the suitability of viticulture heartlands like Bordeaux for producing wine is predicted to fade, while new areas at the coolest edges of the European wine map (including England and even Sweden) could see vines flourish [27, 30]. Across the Atlantic, climate change will similarly reshape wine growing, with more southerly states of the US becoming less suitable [26] and, alongside heat and water stress risks, facing a growing threat from wildfires [28].

In 2017 more than 100 growers in Chile's Central Valley region saw their vineyards damaged or destroyed by fire [31]. Later that same year

California experienced its most destructive wildfire season on record at the time. Their 2018 season was even worse. Around 8,000 fires burned their way across huge areas of land, causing billions of dollars of damage and claiming the lives of over 80 people [32]. Many of the well-irrigated vineyards of northern California were able to swerve direct destruction by the flames in 2017 and 2018, but the smoke that shrouded much of the state meant that grapes, and any wine made from them, risked being tainted. This smoke-taint—where the wine ends up with distinctly unpleasant notes of ashtray—has become a costly side effect of wildfires for many vineyards. Smoke damage from the 2003 bush fires in Australia is estimated to have cost over $4 million [33].

The future of fizz could therefore be one of changing tastes as well as uncertain supplies, but the world's wine growers, especially those in Champagne, are already striving to get ahead of the temperature curve.

* * *

Adjusting grape harvest dates to fit with warm or cool years is an adaptation strategy as old as wine making itself. With the strong warming trend in France over the last few decades, the simple response of earlier harvesting has allowed production and quality to be kept high even in the hottest years. As heat wave, drought and disease risks increase in the future though, this is unlikely to be enough. Switching the times of day used for harvesting, as well as the date, can mean workers are protected from heat stress and the grapes themselves are cooler—this means they then degrade more slowly between field and winery [34].

The winery too must adapt to higher temperatures and changing harvests. The higher-sugar content of the grape juice will demand more alcohol-tolerant yeasts, while extra cooling could mean more energy use, costs and emissions. Good fermentation typically needs cool and stable conditions (10–15 degrees Celsius) and so will require on-site renewables, like solar, or the extension of existing cellars to limit any extra energy and carbon costs during heat waves [35].

Back in the vineyards, deliberately delaying the accelerated ripening brought about by climate change is possible through late pruning—this holds back formation of the new season's buds. As the summer progresses and temperatures hot up, allowing more shading from leaves around bunches of grapes, and so protecting them from scorching, can be effective too. Netting is sometimes used to provide shade alongside protection from birds and hail storm damage [36]—in Australia they've even had success with spraying clay-based sunscreens over the fruit and leaves during heat waves [37]. For newer vineyards, vines can be trained to produce taller stems and so lift the grapes further away from the temperature hot spots that often form close to the soil surface [36].

The soil itself can be a powerful climate-smart ally for wine growers. Those in Champagne benefit from the ability of the chalk soils to hold onto water through dry spells and provide the vines with a slow-release reservoir. Restricting the depth of any tillage and incorporating plenty of organic matter can both help boost the water-retaining properties of the soil and enhance its carbon stocks. The use of mulches—such as vine clippings—and cover crops again helps to boost soil carbon, as well as supressing weeds, reducing evaporation and preventing erosion [27, 36].

Where soil moisture levels drop too low, carefully managed irrigation can provide crucial relief for the plants. Widespread use of irrigation, however, will put extra strain on local water resources that are already likely to be under severe stress in times of drought. Too much irrigation can also lead to a build-up of salts in vineyard soils that then becomes damaging to the vines [34]. Down in the Mediterranean, low rainfall and the absence of irrigation has for centuries been compensated for by using the gobelet pruning method, where the vines are grown as free-standing bushes and their leaf area is cut right back to reduce water losses [4].

More pest and disease attacks may mean greater use of insecticides and fungicides, too but for emerging threats like Pierce's disease, there have been encouraging results for biological controls and, in particular, the use of a cocktail of bacteriophages (viruses that consume the invading bacteria) to contain this costly disease [38].

A warming France should at least mean the devastating effects of late frosts recede over time. To stave them off, growers currently use everything from lighting fires between the rows, to gas heaters, vine-top sprinklers and even wind machines (that mix the air and so prevent frost forming). As these can be expensive, phasing them out could help reduce costs and energy use. But the diminishing risks of late frosts are occurring alongside ever-earlier vine bud burst each spring [39], so the complete abandonment of anti-frost measures would be very risky. Such technical and management strategies can certainly buy wine growers time in the face of climate change. Ultimately though, building long-term climate resilience into their vines and the wines they produce will require new planting.

The undulating topography of the Champagne region lends itself well to producing microclimates that buffer the effects of heat waves and droughts. Planting on north-facing slopes can mean vines are spared from damage during the hottest parts of the day, while selecting areas with deeper soils often provides more reliable soil moisture levels. The enforcers of the Champagne appellation may not be amused, but switching to new grapevine clones or root stocks that extend the time taken until grape maturity, improve disease resistance and give better growth under drought conditions can help ensure the new vines are still fit for purpose decades from now [4, 34].

Finally, and certainly something that is vexing the members of Champagne's Appellation Protection Committee, is the option of a wholesale move to new, more climate-appropriate, locations opening up in the north. Across the English Channel, in southern counties like Kent and Hampshire, this is exactly what is happening. In 2015 the leading champagne house Taittinger bought up 69 hectares of prime farmland in Kent [40]. With its own chalky soils and fast-warming maritime climate, the area is becoming a prime site for growing Chardonnay, Pinot Noir and Pinot Meunier grapes—the backbones of champagne. The quality of English sparkling wines (they still can't be called 'Champagne' under EU law) is already regarded as world class [41]. As a long-term adaptation strategy for the centuries-old champagne houses of France, embracing the idea of this vine-growing Entente Cordiale may help ensure they are still producing wonderful fizz for centuries to come.

References

1. wsta.co.uk. *Prosecco Remains Top of the Pops but English Sparkling Wine is Creeping up the Charts.* https://www.wsta.co.uk/press/848-prosecco-remains-top-of-the-pops-but-english-sparkling-wine-is-creeping-up-the-charts (2017).
2. Bailey, P. Global sparkling wine market trends. *Wine Australia.* https://www.awri.com.au/wp-content/uploads/2018/08/Overview-of-the-sparkling-wine-market-in-Australia.pdf (2018).
3. Ritchie, H. Global wine consumption per capita (litres pure alcohol), 2014. *Ourworldindata.org.* https://ourworldindata.org/grapher/wine-consumption-per-person (2018).
4. van Leeuwen, C. & Darriet, P. The impact of climate change on viticulture and wine quality. *J. Wine Econ.* **11**, 150–167 (2016).

5. vintus.com. *Champagne Bollinger.* https://www.vintus.com/producers/champagne-bollinger/ (2019).

6. champagne.fr. *The Champagne Terroir.* https://www.champagne.fr/en/terroir-appellation/champagne-terroir/champagne-terroir-definition (2019).

7. champagne.fr. *Champagne Terroir: A Dual Climate.* https://www.champagne.fr/en/terroir-appellation/champagne-terroir/a-dual-climate (2019).

8. financesonline.com. *Top 10 Most Expensive Champagne Bottles in the World in 2019.* https://financesonline.com/top-10-most-expensive-champagne-bottles-in-the-world/ (2019).

9. Karlsson, P. & Karlsson, B. Want to buy a vineyard in France? Here's how much it costs. *Forbes.* https://www.forbes.com/sites/karlsson/2015/07/07/want-to-buy-a-vineyard-in-france-heres-how-much-it-costs/#60242e88479d (2015).

10. FAO. *Grapes and Wine: Agribusiness Handbook.* Food and Agriculture Organization of the United Nations. http://www.fao.org/docrep/012/al176e/al176e.pdf (2009).

11. guildsomm.com. *Terroir and the Importance of Climate to Winegrape Production.* https://www.guildsomm.com/public_content/features/articles/b/gregory_jones/posts/climate-grapes-and-wine (2015).

12. Petti, L. *et al. Life Cycle Assessment in the Agri-Food Sector,* 123–184 (Springer, 2015).

13. Doman, E. *7 Most Common Grapevine Diseases.* Wine Cooler Direct. https://learn.winecoolerdirect.com/common-grapevine-diseases/ (2015).

14. Henry, C. 'Fatal' frost hits Champagne vineyards. *Decanter.com.* https://www.decanter.com/wine-news/fatal-frost-hits-champagne-vineyards-367272/ (2017).

15. Burgess, L. Knowing the perfect time to harvest grapes is the difference between good and great wine. *vinepair.com.* https://vinepair.com/articles/knowing-the-perfect-time-to-harvest-grapes-is-the-difference-between-good-and-great-wine/ (2016).

16. champagne411.com. *The Industrial Revolution—Champagne History.* http://www.champagne411.com/champagne-history/industrial-revolution.html (2016).

17. Rugani, B., Vázquez-Rowe, I., Benedetto, G. & Benetto, E. A comprehensive review of carbon footprint analysis as an extended environmental indicator in the wine sector. *J. Clean. Prod.* **54**, 61–77 (2013).

18. WRAP. The life cycle emissions of wine imported to the UK. *Waste & Resources Action Programme.* http://www.wrap.org.uk/sites/files/wrap/The Life Cycle Emissions of Wine Imported to the UK Final Report.pdf (2007).

19. WRAP. Household food and drink waste in the United Kingdom 2012. *Waste and Resource Action Programme.* http://www.wrap.org.uk/sites/files/wrap/hhfdw-2012-main.pdf.pdf (2013).

20. Bono, A. D., Peduzzi, P., Kluser, S. & Giuliani, G. Impacts of summer 2003 heat wave in Europe. *UNEP Environment Alert Bulletin.* http://www.grid. unep.ch/products/3_Reports/ew_heat_wave.en.pdf (2004).

21. García-Herrera, R., Díaz, J., Trigo, R. M., Luterbacher, J. & Fischer, E. M. A review of the European summer heat wave of 2003. *Crit. Rev. Environ. Sci. Technol.* **40**, 267–306 (2010).

22. Mercer, C. Heatwave in French vineyards evokes memories of 2003. *Decanter. com.* https://www.decanter.com/wine-news/heatwave-in-french-vineyards-evokes-memories-of-2003-265451/ (2015).

23. Cook, B. I. Global warming is pushing wine harvests earlier—But not necessarily for the better. *theconversation.com.* https://theconversation.com/ global-warming-is-pushing-wine-harvests-earlier-but-not-necessarily-for-the-better-56407 (2016).

24. EAC. *Heatwaves: Adapting to Climate Change.* House of Commons Environmental Audit Committee. https://publications.parliament.uk/pa/ cm201719/cmselect/cmenvaud/826/826.pdf (2018).

25. Kirk, T. Chardonnay and Pinot Noir under threat from climate change. *The Telegraph.* https://www.telegraph.co.uk/foodanddrink/foodanddrinknews/ 11315083/Chardonnay-and-Pinot-Noir-under-threat-from-climate-change. html (2014).

26. Miller, C. Pinot Noir is wine's polar bear. *slate.com.* https://slate.com/tech-nology/2014/12/wine-and-climate-change-pinot-noir-is-the-vintners-polar-bear.html (2014).

27. Fraga, H., Malheiro, A. C., Moutinho-Pereira, J. & Santos, J. A. An overview of climate change impacts on European viticulture. *Food Energy Secur.* **1**, 94–110 (2012).

28. De Orduna, R. M. Climate change associated effects on grape and wine quality and production. *Food Res. Int.* **43**, 1844–1855 (2010).

29. Cross, A. Best loved wines at risk from climate change. *BBC Online.* https:// www.bbc.co.uk/news/science-environment-11573553 (2010).

30. Hannah, L. *et al.* Climate change, wine, and conservation. *Proc. Natl. Acad. Sci.* **110**, 6907–6912 (2013).

31. Balter, E. After fires, Chile rebuilds devastated wine regions. *Wine Spectator.* https://www.winespectator.com/webfeature/show/id/Chile-Rebuilds-Devastated-Wine-Regions (2017).

32. fire.ca.gov. *Top 20 Most Destructive California wildfires.* http://www.fire. ca.gov/communications/downloads/fact_sheets/Top20_Destruction.pdf (2018).

33. Rogers, A. How climate change and 'smoke taint' could kill Napa wine. *Wired.* https://www.wired.com/story/smoke-taint-napa/ (2017).

34. Neethling, E. *et al. Adapting Viticulture to Climate Change.* ADVICLIM, European Union. http://www.adviclim.eu/wp-content/uploads/2015/06/ B1-deliverable.pdf (2016).

35. Mozell, M. R. & Thach, L. The impact of climate change on the global wine industry: Challenges & solutions. *Wine Econ. Policy* **3**, 81–89 (2014).
36. Sabir, A. *et al.* Sustainable viticulture practices on the face of climate change. *Agri. Res. Technol.* **17**. https://juniperpublishers.com/artoaj/pdf/ARTOAJ. MS.ID.556033.pdf (2018).
37. Riley, L. Sunscreen for winegrapes: Demonstration trial. *Wine Australia.* https://www.wineaustralia.com/getmedia/77df1eb2-63f8-491b-aaa5-783461ceb7a4/201411-Sunscreen-for-winegrapes.pdf (2014).
38. Das, M., Bhowmick, T. S., Ahern, S. J., Young, R. & Gonzalez, C. F. Control of Pierce's disease by phage. *PLoS One* **10**, e0128902 (2015).
39. Molitor, D. *et al.* Late frost damage risk for viticulture under future climate conditions: a case study for the Luxembourgish winegrowing region. *Aust. J. Grape Wine Res.* **20**, 160–168 (2014).
40. BBC. *Champagne House Taittinger Buys Kent Apple Orchard.* https://www. bbc.co.uk/news/uk-england-kent-35060681 (2015).
41. Decanter.com. *England.* https://www.decanter.com/tag/english-wine/ (2019).

CHAPTER 15

Conclusion

Abstract Just one day's food and drink has together travelled over 40,000 miles and encompassed a wide array of climate change risks and responses. Their carbon footprints also ranged widely, with milk, chicken and fried food standing out as the biggest overall, but with tea and coffee having very high emissions for each gram actually consumed. On-farm emissions and those from storage and cooking dominate life-cycle emissions for most foods. All face threats from climate change, with many smallholders around the world facing the prospect of losing their livelihoods altogether as impacts intensify. Every food and drink examined, however, has a raft of climate-smart responses available that could boost resilience and yields, and lower emissions. Engaging with stakeholders, especially the people who produce our food, and taking local contexts into account when designing and applying solutions for food in a changing climate is the fundamental take-home message.

Keywords Food miles • Vegan • Vegetarian • Mycoprotein • Soya • Insect flour • Patrick Geddes • Carbon footprint

Our day's food, albeit an extravagant champagne-swilling one, has taken in 5 continents and myriad nations of the world. Along the way it has borne witness to a warming world where severe weather events are

© The Author(s) 2019
D. Reay, *Climate-Smart Food*,
https://doi.org/10.1007/978-3-030-18206-9_15

becoming more frequent and intense, and where the supply chains that feed us are buckling. From impacts that reach across whole continents, like retreating Arctic ice and faltering Himalayan melt waters, to the up-close-and-personal threats of fire, pests and disease, there's a taste of climate change in every bite.

In getting from source to our West Lothian kitchen, the day's meals have together clocked up a staggering 40,000 miles of travel (mainly by container ship[1]). Such huge 'food miles' certainly play their part in the total carbon footprint,[2] but it is actually only a minor one for most. As long as airfreight is avoided, the bulk of the life-cycle emissions occur back on the farms or during storage and cooking. Milk stands out as a major slice of our carbon pie (methane belched by cows being the main culprit) (Fig. 15.1). The chicken curry, fish supper and champagne are big players too, between them making up nearly two-thirds of the total. At least for the latter two, I can guarantee neither a scrap of batter nor a drop of fizz is ever wasted.

For many of us it is the amount of meat and dairy products we consume that dominates the carbon footprint of our food. Since the early 1960s the amount of meat produced worldwide has risen almost fivefold—the average person in the UK now consumes over 75 kilograms per year of it (in the US, the average is over 100 kilograms). This gravy tide may be turning though.

Two-thirds of US citizens say they are now eating less of at least one type of meat, and a third of Brits say they have either cut back or given up meat altogether [1]. My own family's diet has itself changed quite radically as the writing of this book has progressed. At its inception we still ate some bacon, sausages and salami, alongside fresh beef, lamb, pork and chicken. First to go was the processed meat as more and more studies

[1] 40,828 miles based on container ship to Portsmouth, England, from Admiral Barroso Terminal (orange juice), Kolkata (tea), Djibouti (coffee), Takoradi (cocoa), Vera Cruz (maize), Karachi (rice), Reykjavik (cod), Calais (champagne) and onward delivery to Scotland by truck.

[2] 4.06 kilograms carbon dioxide equivalent (CO_2e), based on consumption of one person: 200 ml glass of orange juice (100 g CO_2e), 200 ml mug of black tea (50 g CO_2e), 200 ml glass of milk (600 g CO_2e), 40 g chocolate bar (270 g CO_2e), 200 ml mug of black coffee (150 g CO_2e), banana (200 g CO_2e), portion of chicken curry (75 g) with rice (100 g) (525 g CO_2e), 40 g bag of nachos (50 g CO_2e), standard portion fish & chips (1,000 g CO_2e), three 125 ml glasses (half bottle) of champagne (1,000 g CO_2e).

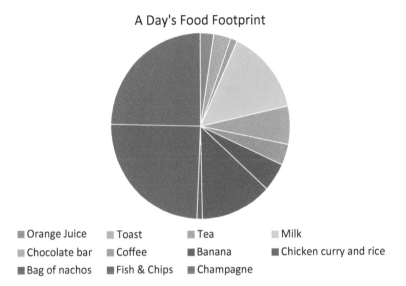

Fig. 15.1 Relative contribution of the day's foods and drinks to the total carbon footprint of 4 kilograms. For details see footnote 2

emerged on the cancer risks they pose [2]. Next went the ruminants—the beef and lamb—due to their high carbon footprints (up to 4 kilograms of emissions per serving of lamb and even more than that for beef) [3].

An initial dabble with meat substitutes like mycoprotein Quorn sausages and soya-filled Linda McCartney pies was a big success and saw the end of our pork consumption too. Chicken dinners held out for a while longer (until the last barbecue of summer 2018). Chicken has a carbon footprint only slightly higher than that of meat substitutes like Quorn [4], and by always buying free range we assumed the welfare of the birds was guaranteed. Researching broiler chickens for this book (Chap. 9) made clear my imagined daisy-pecking idyll for commercial free-range birds is often far from reality. We still eat lots of eggs though, as our small flock of rescue hens in the back garden provide enough for us and many of our friends and neighbours.

Most recently we have switched to plant-based milk—dairy milk clocks up around 600 grams of emissions per serving, while the plant-based ones cut this by two-thirds [3]. Crucially, it tastes good too. Over the past couple of years, the carbon footprint of our food has contracted along with my own waistline. Cutting out meat has made the biggest dent on both counts. Our family diet continues to evolve as the kids become more food and climate-aware, and much more vociferous in their opinions. The 6-year-old Molly who liked to wrap her cocktail sausages in salami is now the 12-year-old extolling the virtues of vegan cheese. We are yet to incorporate insects into our daily diet—these already form part of the traditional diet for around 2 billion people worldwide [5] and could be a viable lower-carbon alternative to animal protein in Western diets too [6]. Likewise, Impossible Burgers (a plant-based burger using heme proteins from legumes) [7], Greggs' vegan sausage rolls (they had all sold out), and artificial steaks (from lab-grown animal cells) [8] remain culinary treats for the future.

Cricket flour may still be some years away from appearing on the shelves of our own local food store, but other alternatives to meat and dairy have become cheaper and more readily available. Whatever our food choices, they are certainly becoming better informed. From quick and easy food and climate comparison tools [3] to in-depth reports on the sustainability of our diets [9, 10], we have never been better able to assess the relative merits of what's in our shopping basket. The message is a clear one: a transition to more sustainable diets can deliver big benefits for our own health alongside that of the planet—an estimated 10 million lives saved each year, just for starters [9].

Yes, there are barriers, risks and uncertainties aplenty. Wholesale conversion to plant-based milk, for instance, could cripple the dairy industry and damage the livelihoods of already-struggling farmers. The nutritional benefits of meat and milk substitutes will differ from their animal-based cousins [4] and soaring demand for soya and almond milk could accelerate land-use change and deforestation overseas, thus shifting some emissions and biodiversity loss problems offshore [11].

For much of the world, uncertain projections of our future climate are a barrier too. Most remain far too broadscale and imprecise to be useful for individual farmers to decide exactly what to grow and when. Likewise, specially-bred crop varieties have huge potential, but developing these takes time and, even when available, farmers may be distrustful of them [12].

If researching this book has taught me anything, it is that climate-smart food is not a one-size-fits-all solution. Climate change is a global phenomenon, but its impacts on our food, and the best responses, are more diverse and location specific even than the rainbow of Mexican maize landraces. The further we as consumers are separated from the farmers and herders, the vintners and fisherfolk, the less we are able to understand this complexity, and the greater the chance that vital local contexts are overlooked.

At the risk of making my Scottish sustainability hero Patrick Geddes turn in his grave, we need to 'Think Local, Act Global' on food. To learn from and enhance the capacity to face climate change at the local level, everywhere. Through greater support for and engagement with the many millions of people who help feed us each day, the opportunity to realise a climate-smart future for our food is within reach. Let's grasp it.

REFERENCES

1. Ritchie, H. Which countries eat the most meat? *BBC Online.* https://www.bbc.co.uk/news/amp/health-47057341 (2019).
2. Gallagher, J. Processed meats do cause cancer—WHO. *BBC Online.* https://www.bbc.co.uk/news/health-34615621 (2015).
3. Stylianou, N., Guibourg, C. & Briggs, H. Climate change food calculator: What's your diet's carbon footprint? *BBC Online.* https://www.bbc.co.uk/news/science-environment-46459714 (2018).
4. Ritchie, H., Reay, D. S. & Higgins, P. Potential of meat substitutes for climate change mitigation and improved human health in high-income markets. *Front. Sustain. Food Syst.* **2**, 16 (2018).
5. Payne, C. Entomophagy: How giving up meat and eating bugs can help save the planet. *The Independent.* https://www.independent.co.uk/news/long_reads/entomophagy-eat-insects-food-diet-save-planet-meat-cattle-deforestation-a8259991.html (2018).
6. Akhtar, Y. & Isman, M. *Proteins in Food Processing* 263–288 (Elsevier, 2018).
7. Sear, J. *What Does the Impossible Burger Taste Like?* https://www.bbcgoodfood.com/howto/guide/what-does-impossible-burger-taste (2017).
8. Simon, M. Lab-Grown meat is coming, whether you like it or not. *Wired.* https://www.wired.com/story/lab-grown-meat/ (2018).
9. Willett, W. *et al.* Food in the Anthropocene: The EAT–Lancet Commission on healthy diets from sustainable food systems. *Lancet* **393**, 447–492 (2019).
10. FCRN. *Food and Climate Research Network.* https://fcrn.org.uk (2019).

11. Franklin, E. Milk: The sustainability issue. *sustainablefoodtrust.org.* https://sustainablefoodtrust.org/articles/milk-the-sustainability-issue/ (2017).
12. Rosenstock, T. S., Nowak, A. & Girvetz, E. *The Climate-Smart Agriculture Papers Investigating the Business of a Productive, Resilient and Low Emission Future* (Springer, 2019).

INDEX[1]

[1] Note: Page numbers followed by 'n' refer to notes.

CPSIA information can be obtained
at www.ICGtesting.com
Printed in the USA
LVHW070156150819
627728LV00014B/85/P

9 783030 182052